Elementary Solutions to Longstanding and Contemporary Scientific Problems

Julio A. Gonzalo

C y C

Madrid, 2018

Elementary Solutions to Longstanding and Contemporary Scientific Problems

1st Edition May 2018

© Julio A. Gonzalo

© Asociación Española Ciencia y Cultura
c/ Pavía 4, 1º D. 28013 Madrid (España)
Fax: (34) 91-4978579
www.cienciaycultura.com
E-mail: julio.gonzalogonzalez@ gmail.com
E-mail : aecienciaycultura@gmail.com

ISBN: 978-17-188892-6-2

CONTENTS

Prologue 4
1. Selected Mathematical Problems 5
Chapter 1.1 Introductory Considerations 6
Chapter 1.2 On Fermat's Last Theorem 10
Chapter 1.3 On Golbach's Conjecture 14
Chapter 1.4 On Twin Primes 19
2. Selected Statistical Problems 25
Chapter 2.1 Introductory Considerations 26
Chapter 2.2 On the Critical Isotherms in Ising Systems. 32
Chapter 2.3 On the 3D Ising Critical Exponents. 40
Chapter 2.4 On the Equation of State for 3D Ising Systems. 55
Chapter 2.5 Effective Critical Exponents at Minimal dimensionalities. 66
Chapter 2.6 Ferroelectric Dipole Waves. 79
3. Selected Global Problems 94
Chapter 3.1 Introductory Considerations 95
Chapter 3.2 On a Rate Equations Approach to World Population Trends 97
Chapter 3.3 On World Population Slowing Down 116
Chapter 3.4 At which side will be history at the end of the 21st Century? 128

Prologue

In the present book several selected papers, some not yet published, some previously published, on longstanding or recent mathematical, statistical and global problems are put together. The reader may judge by himself whether the elementary proofs or general considerations put forward in them are fully conclusive or not and to what extent.

Introductory considerations for non-specialists are given in Sections 1.1, 2.1 and 3.1.

I hope that the selected problems are truly interesting for a potentially wide readership with a reasonably good scientific background.

Helpful interactions with my good friends and colleagues Manuel Alfonseca, Ginés Lifante, Pedro Tarazona, Carmen Aragó, and Manuel I. Marqués, are gratefully acknowledged.

Julio A. Gonzalo
May, 2018

1. Selected Mathematical Problems

1.1 Introductory Considerations

1.2 On Fermat's Last Theorem

1.3 On Golbach's Conjecture

1.4 On Twin Primes

CHAPTER 1.1
INTRODUCTORY CONSIDERATIONS

The case of Fermat's last theorem is very famous. Pierre Fermat (1601-1665) one of the greatest French mathematicians of all times had left written a note on the margin of one of his handbooks saying that he had a proof that the equation

$x^n + y^n = z^n$ (x, y, z, n all integers)

had no solutions for $n \geq 3$. The note said that he had not enough space to write down the proof in the small space of that margin, but it seemed to imply, nonetheless, that it was a relatively short one. Fermat was a professional lawyer, but his contributions to create analytic geometry and probability calculus were extraordinary and, since his death, many amateur and professional mathematicians have tried, unsuccessfully, to prove that the above simple relationship had no solutions for $n \geq 3$ integer.

In 1993 Andrew Wiley, a British Professor at Oxford, later Sir Andrew Wiley, F.R.S., caused sensation all around the world when he announced that he had finally demonstrated Fermat's Last Theorem. Something was found incorrect in Wiley's original proof, but he corrected it, and his 1995 paper on the subject was generally accepted ad satisfactory. His proof, very long and unlikely to be anything related to Fermat's short tentative proof, was based upon the "modularity theorem" which involves elliptic curves. This theorem was unproven around 1986, and was known at that time as the Taniyama-Shimura-Weil conjecture. At that time most mathematicians believed that this conjecture was "impossible to prove". Wiley dedicated all his available time to prove Fermat's last theorem without telling anybody about it except his wife. In 1994 Wiley did publish a second paper in which he circumvented the flaw in his first paper, in collaboration with his former graduate student Richard Taylor; Mathematicians all around the world accepted as valid the long and elaborated proof of Fermat's last theorem.

In section 1.2 it will be shown that taking into account that for $n = 1$

$(a - b) + 2b = (a + b)$ has infinite solutions,

that for $n = 2$

$(a^2 - b^2)^2 + 4a^2b^2 = (a^2 + b^2)^2$ has infinite solutions (**less numerous**)

and that for $n = 3$

$(a^3 - b^3)^3 + y^3 = (a^3 + b^3)^3$ has **no** solutions in y = integer, because $y = [2b^3(3a^6 + b^6)]^{1/3}$ = irrational for $a > 1$ and $b < a$, solutions in integers for $n \geq 3$ should not be expected, in principle. Are the elementary considerations given in Section 1.2 satisfactory as a proof of Fermat's last theorem? The reader should judge for himself.

***

The case of Goldbach's conjecture is one of the oldest unsolved problems of number theory. Christian Goldbach (1690-1764), in a letter (1742) to Leonhard Euler, one of the greatest mathematicians of all times, proposed his unproven assumption that every **even** number could be given as the sum of two prime numbers. Using computers, Golbach conjecture has been verified up to $n \leq 4 \times 10^{18}$, and everybody expects that even for much larger even numbers the conjecture will be still valid, in spite of the fact that the fraction of prime numbers decreases (relatively slowly, but consistently) as n grows.

In 1923 Hardy & Littlewood conjectured that the number of representations of a large integer n as a function of the sum of c primes $n = p_1 + p_2 + \ldots p_c$, with $p_1 < p_2 < \ldots p_c$ for the case of $c = 2$ should be equal to

$$\Pi_2 = \prod_{p \geq 3}\left(1 - \frac{1}{(p-1)^3}\right) = 0.6601618\ldots$$

which is known as the Hardy-Littlewood twin prime constant.

In Section 1.3 unpublished work by Alfonseca and Gonzalo will be shown to result in $n_{\min}(N) = C_2 N / (\ln N)^2$ as the **minimum** number of ways in which N = even can be given as the sum of two primes and $n_{\max}(N) = 3.6\, C_2 N / (\ln N)^2$ as the corresponding **maximum** number, where $C_2 = 0.6601618\ldots$ This shows that the number of ways in which Goldbach's conjecture is fulfilled oscillates widely as N increases, but that is always further from zero as N increases indefinitely.

As a **corollary** of the fact that the number of twin primes $[(n-1), (n+1)]$ (to be proven in Sec. 1.4) grows indefinitely towards infinity, it may be stated that there are indefinitely large **even** numbers, going toward infinity, which can be represented as the sum of two twin primes: $(n-1) + (n+1) = 2n$, which is **directly related** to Goldbach's conjecture.

The famous proof by reduction ad absurdum that the number of prime numbers (integers n such that they have only two divisors $\leq n$, including 1 and n) is infinity is known from the time of Euclid (330 a. C., 275 a. C). It is well known also that according to the prime number theorem, due to Legendre and Gauss

$$\lim_{x\to\infty} \frac{\Pi(x)}{x/\ln x} = 1 ,$$

where $\Pi(x)$ is the total number of primes less than or equal to x and x is a real positive number.

On the other hand, since the time of Euclid, it is suspected that the number of twin primes, like (5, 7), (11, 13) etc. is also infinite, but a conclusive proof of this, apparently, is lacking. Section 1.4 reproduces elementary considerations which seem to provide a conclusive proof that the number of twin primes $(n-1, n+1)$ grows indefinitely without any upper limit while becoming increasingly more rare as x grows.

Pierre Fermat (1601-1665)

Chapter 1.2
On Fermat's Last Theorem

In 1993 Professor Andrew Wiles reported success in proving Fermat's Last Theorem in a 200 pp. paper after a seven year quest. Here it is shown than elementary quantitative considerations may be sufficient to prove that $x^n + y^n = z^n$ has no solutions in integers for $n \geq 3$.

ELEMENTARY SOLUTIONS

Fermat's Last Theorem, states that $x^n + y^n = z^n$, where (x, y, z) are **triplets** of positive integers, has no solutions for $n \geq 3$, a positive integer.

Around the year 2000 B.C. the Babylonians sought already a way to break down an square number into a sum of two squares. The great Greek mathematician Pythagoras, as it is well known, was able then to prove his famous theorem. His demonstration implies conclusively that Fermat's Last Theorem has infinite solutions in integers (x, y, z) for $n = 2$.

In the seventeenth century the great French mathematician Pierre Fermat wrote that while the square of a whole number can be easily broken into two squares of whole numbers (for instance $5^2 = 4^2 + 3^2$) the same cannot be done with cubes or any higher powers. After Fermat's death many professional and amateur mathematicians have devoted their time, unsuccessfully, to prove that theorem.

Let us investigate the possibility of doing an elementary quantitative proof of Fermat's Last Theorem.

First we prove that $N_m(z_m)$, the number of **triplets** (x, y, z) with $z \leq z_m$ decreases as **n** increases for $n = 1, 2, 3...$

In particular **n = 1**:

$(a - b) + 2b = (a + b)$ →

$$(2, 1)$$
$$(3, 2), (3, 1)$$
$$\text{------}$$
$$(a_m, a_m - 1), (a_m, a_m - 2)...(a_m, 1)$$

resulting in

$$N_1(z_m) = \frac{(z_m-1)(z_m-1)}{2} \xrightarrow[z_m \gg 1]{} \log_{z_m} N_1(z_m) \simeq 2$$

For **n = 2**:

$(a^2 - b^2) + 4a^2b^2 = (a^2 + b^2)^2 \rightarrow$

$$((2^2)^2, (1^2)^2)$$
$$((3^2)^2, (2^2)^2), ((3^2)^2, (1^2)^2)$$
$$\text{------}$$
$$((a_m^2)^2, \left[(a_m-1)^2\right]^2), \ldots ((a_m^2)^2, \left[(1)^2\right]^2)$$

resulting in

$$N_1(z_m) = \frac{(z_m^{1/2}-1)(z_m^{1/2}-1)}{2} \xrightarrow[z_m \gg 1]{} \log_{z_m} N_2(z_m) \simeq 1$$

For **n = 3**:

It is well known that the equation $x^3 + y^3 = z^3$, $xyz \neq 0$ has no solution in integers (See for instance Javier Cilleruelo and Antonio Córdoba: "La Teoría de los Números", Biblioteca Mondadori, Madrid, 1992).

In particular there are no solutions like

$(a^3-b^3)^3 + [6(a^3)^2b^3 + 2(b^3)^3] = (a^3 + b^3)^3$

because it would require

$[6(a^3)^2b^3 + 2(b^3)^3] = [3a^6 + b^6]2b^3 = y^3$,

which implies

$$[3a^6 + b^6] = (2b^3)^3,$$ and therefore $a = b$, which makes $x^3 = 0$.

Therefore $N_n(z_m)$ decreases consistently for n decreasing $n = 1$, $n = 2...$, as it might be expected, and we can plot $\log_{z_m} N_n(z_m)$ vs n as in Fig 1.

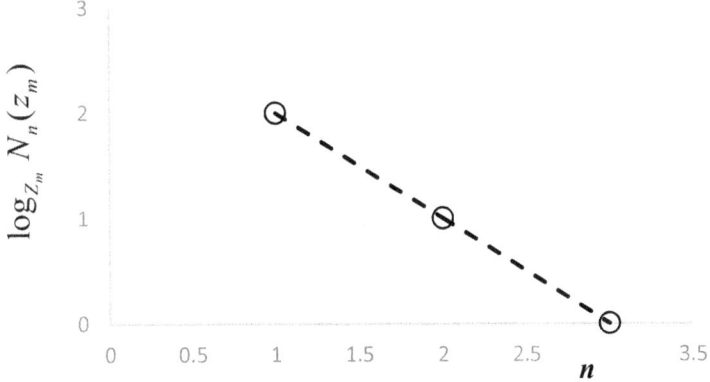

Fig 1 shows directly that we cannot expect solutions of $x^n + y^n = z^n$ in integers for $xyz \neq 0$, $n \geq 3$ integer if it is granted that $N_n(z_m)$ decreases for increasing n.

The possibility that Fermat had perhaps in mind a short argument along this line to prove his famous Last Theorem should not be discarded.

Christian Goldbach (1690-1764)

CHAPTER 1.3
ON GOLBACH'S CONJECTURE

Goldbach's conjecture says that any even number can be decomposed in one or more ways into the sum of two primes: $N(\text{even}) = p + p'$. The number of ways in which N

can be decomposed in such a way, $n(N)$, grows with N as can be checked empirically. Unpublished work by Alfonseca and Gonzalo shows that $n(N)$ oscillates between $n(N)_{min} = C_2 N / (\ln N)^2$ and $n(N)_{max} \simeq 3.6 \cdot C_2 N / (\ln N)^2$ where $C_2 = 0.6601618...$ is the twin prime number. In what follows it is argued that $n(N)$ is directly related to the total number of twin prime pairs less than N by elementary argumentation.

Goldbach's conjecture, $N(\text{even}) = p + p'$ with p, p' primes, has defied rigorous proof for centuries and it has attracted the attention of amateur and professional[1-4] mathematicians since it was formulated by Goldbach in 1742.

Any even integer N can be decomposed into sums such as

$$N = (N-3) + 3, (N-5) + 5, \ldots (N-p) + p, \ldots (N - p_{max}) + p_{max} \qquad (1)$$

where, according to the prime number theorem the number of prims equal or below p_{max} is given by $\Pi(N)$ such that

$$\lim_{N \to \infty} \frac{\Pi(N)}{N / \ln N} = 1 \qquad (2)$$

We can define

$$C(N) \equiv \frac{\Pi(N)}{N/\ln N} \quad (3)$$

and use the following approximation for $N \gg 1$,

$$\left(\frac{\Pi\left(\frac{1}{2}N-p\right)\Pi\left(\frac{1}{2}N+p\right)}{\frac{1}{2}N-p\quad\frac{1}{2}N+p}\right)^{1/2} \simeq \left(\frac{\Pi\left(\frac{1}{2}N-1\right)\Pi\left(\frac{1}{2}N+1\right)}{\frac{1}{2}N-1\quad\frac{1}{2}N+1}\right)^{1/2} \simeq C\left(\frac{1}{2}N\right)\frac{1}{\ln\frac{1}{2}N} \quad (4)$$

Then we can approximate n (N), the number of ways in which N can be decomposed into the sum of two primes, by means of

$$n(N) = \sum_p \left\{ \left(\frac{\Pi\left(\frac{1}{2}N-p\right)}{\frac{1}{2}N-p} \right) \left(\frac{\Pi\left(\frac{1}{2}N+p\right)}{\frac{1}{2}N+p} \right) + \left(1 - \frac{\Pi\left(\frac{1}{2}N-p\right)}{\frac{1}{2}N-p} \right) \left(\frac{\Pi\left(\frac{1}{2}N+p\right)}{\frac{1}{2}N+p} \right) + \left(\frac{\Pi\left(\frac{1}{2}N-p\right)}{\frac{1}{2}N-p} \right) \left(1 - \frac{\Pi\left(\frac{1}{2}N+p\right)}{\frac{1}{2}N+p} \right) + 2 \left(\frac{\Pi\left(\frac{1}{2}N-p\right)}{\frac{1}{2}N-p} \right) \left(\frac{\Pi\left(\frac{1}{2}N+p\right)}{\frac{1}{2}N+p} \right) \right\}$$

$$\cong \Pi(N) \left\{ 2\left(\frac{\Pi\left(\frac{1}{2}N-1\right)}{\frac{1}{2}N-1}\right)\left(\frac{\Pi\left(\frac{1}{2}N+1\right)}{\frac{1}{2}N+1}\right) + \left(1-\frac{\Pi\left(\frac{1}{2}N-1\right)}{\frac{1}{2}N-1}\right)\left(\frac{\Pi\left(\frac{1}{2}N+1\right)}{\frac{1}{2}N+1}\right) + \left(\frac{\Pi\left(\frac{1}{2}N-1\right)}{\frac{1}{2}N-1}\right)\left(1-\frac{\Pi\left(\frac{1}{2}N+1\right)}{\frac{1}{2}N+1}\right) + 2\left(\frac{\Pi\left(\frac{1}{2}N-1\right)}{\frac{1}{2}N-1}\right)\left(\frac{\Pi\left(\frac{1}{2}N+1\right)}{\frac{1}{2}N+1}\right) \right\}$$

$$\cong \left[C(N) \frac{N}{\ln N} \right] \left[C\left(\frac{1}{2}N\right) \frac{1}{\ln \frac{1}{2}N} \right] = A(N) \frac{N}{(\ln N)^2}$$

As mentioned in the Abstract of unpublished work by Alfonseca and Gonzalo[5], for N >> 1 $n(N)$ oscilates between

$$n_{max}(N) = A_{max} \frac{N}{(\ln N)^2} \text{ with } A_{max} \approx 3.6 C_2 \qquad (6)$$

Where $C_2 = 0.6601618\ldots$ is the twin prime number.

It may be noted that $C(N) = C_2 e^{1/\ln N}$ and that for $N \to \infty$, $C(\infty) = C_2$.

REFERENCES

[1] Wang Yuan Ed. *Goldbach conjecture* (World Scientific: Singapore 1984)
[2] I. M. Vinogradow, Comptes Rendu (Doklady) de l'Academie des Sciences de l'URSS, 1937. Vol. XV, Numbers 6 – 7.
[3] Wang Yuan, *On the representation of large even numbers as a sum of two almost primes*, Science Record, New Ser. Vol, No. 5,1957.
[4] Chen Jing – Run, *On the representation of a larger even integer as the sum of a prime and the product of at most two primes*.
[5] Scientia Sinica, Vol. XVI, No. 2, May, 1973.
[6] Manuel Alfonseca, Julio A. Gonzalo, *An empirical study of the number of ways Goldbach conjecture is fulfilled* (not published).

CHAPTER 1.4
ON TWIN PRIMES

The number of twin primes ($x-1$, $x+1$), $x \equiv$ even, is shown to be increasingly well approximated by $\Pi_t(x) = \left(\dfrac{a(x)}{\ln x}\right)\dfrac{x}{\ln x}$ where $a(x)$ approaches a constant $a(\infty)$ for $x \to \infty$. Numerical data for increasing x show that $a(\infty) \to 2.7095$ for $x = 10^{18}$ and decreases consistently for increasing x. Therefore, as expected, the total number of twin primes is infinite because $a(x) \cdot x$ grows faster than $(\ln x)^2$. $a(x)$ is shown to be increasingly well approximated by $a^*(x) = a^*(\infty)e^{1/\ln x}$ with $a^*(\infty) = 4C_2 = 2.64064...$ with C_2 given by the twin prime constant as given by the Hardy and Littlewood conjecture.

According to the prime number theorem[1]

$$\lim_{x\to\infty}\frac{\Pi(x)}{x/\ln x}=1 \qquad (1)$$

where $\Pi(x)$ is the total number of primes less than or equal to x and x is a real positive number.

$\Pi(x)$ is made up of all single primes, like (23,), (,37), (,47) etc., and all twin primes, like (5,7), (11,13), (17,19), etc.

We can write

$$\Pi_s(x) \equiv \alpha_s(x)\Pi(x) \qquad (2)$$

$$\Pi_t(x) \equiv \alpha_t(x)\Pi(x) \qquad (3)$$

Let us define $a(x)$ as

$$a(x) \equiv \frac{\Pi(x)}{x}\ln x \qquad (4)$$

Having into account the prime number theorem, Eq. (1), one can write for x sufficiently large

$$\frac{\Pi(x-1)}{x-1} \approx \frac{\Pi(x+1)}{x+1} \approx a(x)\frac{1}{\ln x} \qquad (5)$$

where we are interested in the behavior of $a(x)$ for $x \to \infty$.

Then, probabilistic considerations suggest that

$$\alpha_s(x) \approx \frac{\left(\frac{\Pi(x-1)}{x-1}\right)\left(1-\frac{\Pi(x+1)}{x+1}\right) + \left(1-\frac{\Pi(x-1)}{x-1}\right)\left(\frac{\Pi(x+1)}{x+1}\right)}{\left\{\left(\frac{\Pi(x-1)}{x-1}\right)\left(1-\frac{\Pi(x+1)}{x+1}\right) + \left(1-\frac{\Pi(x-1)}{x-1}\right)\left(\frac{\Pi(x+1)}{x+1}\right) + 2\left(\frac{\Pi(x-1)}{x-1}\right)\left(\frac{\Pi(x+1)}{x+1}\right)\right\}}$$

$$\approx 1 - a(x)\frac{1}{\ln x} \qquad (6)$$

$$\alpha_t(x) \approx \frac{2\left(\frac{\Pi(x-1)}{x-1}\right)\left(\frac{\Pi(x+1)}{x+1}\right)}{\left\{\left(\frac{\Pi(x-1)}{x-1}\right)\left(1-\frac{\Pi(x+1)}{x+1}\right) + \left(1-\frac{\Pi(x-1)}{x-1}\right)\left(\frac{\Pi(x+1)}{x+1}\right) + 2\left(\frac{\Pi(x-1)}{x-1}\right)\left(\frac{\Pi(x+1)}{x+1}\right)\right\}}$$

$$\approx a(x)\frac{1}{\ln x} \qquad (7)$$

Let us define further

$$a^*(x) \equiv C(x)f(1/\ln x) \qquad (8)$$

with $C(x)$ equal to an x dependent prefactor such that $a^*(x) = a(x)$ for $x \to \infty$, and investigate how does that this work for $f(1/\ln x) = e^{1/\ln x}$
i.e., for

$$a^*(x) = C(x)e^{1/\ln x} = C(x)\left[1 + \left(\frac{1}{\ln x}\right) + \frac{1}{2!}\left(\frac{1}{\ln x}\right)^2 + \ldots + \frac{1}{n!}\left(\frac{1}{\ln x}\right)^n + \ldots\right] \qquad (9)$$

Note that the expansion in Eq. (9) is significant: an expansion in $(\ln x)^n$ instead of $(1/\ln x)^n$ would not work properly for $x \to \infty$.

Table I

x	$\alpha_1(x)$	$a(x)$	$C(x) = a(x)/e^{1/\ln x}$	$C(x) - 4C_2$
10^6	0.2081	2.8751	2.6744	0.0336
10^9	0.1346	2.7913	2.6119	0.0191
10^{12}	0.0994	2.7486	2.6509	0.0102
10^{15}	0.0788	2.7249	2.6959	0.0064
10^{18}	0.0653	2.7095	2.6449	0.0042

Table I above shows that $C(x)$ and $4C_2 = 2.64064\ldots$ approach gradually each other for $x \to \infty$ (C_2 is given by the Hardy and Littlewood conjecture). This can be more clearly seen in Figure 1.

The real values for $a(x)$ fluctuate at low x values, but they come close gradually to the corresponding calculated values for $a^*(x) = a^*(\infty)e^{1/\ln x}$ up to $x = 10^{18}$.

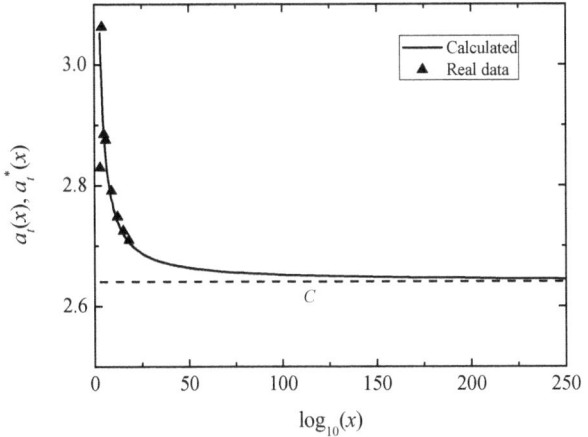

Figure 1.- Plot of $a(x)$ (triangles), as defined in Eq. (4), and $a^*(x)$ (continuous line) as given by Eq. (9), as a function of x. Dashed line corresponds to $C(\infty) = 2.64064... = 4C_2$ with C_2 the twin primes constant as conjectured by Hardy and Littlewood.

Acknowledgements

I would like to thank Ginés Lifante, Manuel Alfonseca and Pedro Tarazona for numerical calculations and competent criticism, as well as to Antonio Córdoba for helpful comments. And to Maggie MacKnee for her interesting work in *Nature* on prime pairs.

REFERENCES

[1] Mc Knee, M. (2013) *First proof that prime numbers pair up to infinity.* Nature, doi: 10.103038/nature2013.12989 and references therein.

[2] Zhang, Y. (2014) *Bonded gaps between primes.* Annals of Mathematics 179 (3) 1121-1174. doi: 104007/annals2014.179.3.7 and references therein.

2. Selected Statistical Problems

2.1 Introductory Considerations.

2.2 On the Critical Isotherms in Ising Systems.

2.3 On the 3D Ising Critical Exponents.

2.4 On the Equation of State for 3D Ising Systems.

2.5 Effective Critical Exponents at Minimal Dimensionalities

2.6 Ferroelectric Dipole Waves.

CHAPTER 2.1
INTRODUCTORY CONSIDERATIONS

The statistical physics of cooperative phenomena and phase transitions in condensed matter has been, one of the main fields of research already since the end of the 19th century, but, very especially, from the second half of last century.

The effective field or mean field approach used by Pierre Weiss at the beginning of the 20th century to theoretically describe ferromagnetic transitions in iron, nickel and cobalt can be considered a first successful approach. It was not clear at the time that the van der Waals theory proposed earlier to describe liquid-vapor phase transitions was also a kind of effective field approach or mean field approach theory analogous to Weiss theory[1] describing ferro-paramagnetic transitions.

The accompanying **Table 2-1** gives a set of phase transitions and registers the author of their first quantitative experimental description as well as the author / authors of the first effective field theory describing satisfactorily, while not exactly, each phase transition

---

[1] J.A. Gonzalo, *Effective field approach to Phase Transitions...* (World Scientific, Singapore, 1991)

Table 2-1 Phase transitions

Transition	Experimental description	Effective field theory
Liquid-Vapor	T. Andrews, 1872	J. van der Waals, 1874
Ferro-Paramagnet	P. Curie, 1895	P. Weiss, 1907
Superconductor Normal	K. Onnes, 1906	Bardeen-Cooper-Schrieffer, 1951
Order/Disorder Alloys	G. Tamman, 1919	W. Bragg, E. Williams, 193
Ferro-Paraelectric	J. Valasek, 1921	W. Mason, 1947
Superfluid-Normal	W. Keesom, A. Keesom, 1935	N. Bogolinbov, 1946
Ferro-Paraelastic	-------	K. Aizu, 1969

Informative monographs introducing modern theories of phase transitions are, among others, the following:

H.E. Stanley, "Introduction to Phase Transitions and Critical Phenomena" (Oxford University Press, Oxford, 1971)

P. Pfeuty and G. Toulouse, "Introduction to the Renormalization Group and to Critical Phenomena" Wiley, New York, 1978)

Shang Keng Ma, "Modern Theory of Critical Phenomena" (W.A. Benjamin Reading, MA. 1976).

C. Domb and M.S. Green (eds). "Phase Transitions and Critical Phenomena" (Academic Press, London, 1972).

An important early development to describe phenomenologically phase transitions was the Landau – Ginzburg theory of phase transitions. A few decades afterwards the pioneering work of K. G. Wilson on the renormalization group provided recursion formulae which put forward after 1971 a wholly new approach to the theory of cooperative phenomena soon rewarded with the Nobel Prize for Physics.

Phase transitions are accompanied by symmetry breakdowns. For example, a ferromagnetic at a temperature T bellow Tc shows a preferred magnetization direction. The whole crystal if it is monodomain, or given domains within the crystal, if it is polidomain, are characterized by a definite magnetization direction. As the temperature increases and arrives to $T = T_C$ unit dipoles within each domain are allowed to point in both opposite directions: the symmetry breaks down. Elementary magnetic moments are disordered, due to thermal energy or, if T_C is very low, due to the quantum zero point energy which can play a comparable to that of the thermal energy for very low T_C's.

Phase transitions can take place **continuously**, in the so called second-order transitions, or **discontinuously**, in the so called first order transitions, as the temperature goes back and forth through T_C.

Phase transitions can be grouped into **universality classes**: a physical system undergoing a phase transition can be made up of one-dimensional **chains** of interacting atoms or molecules of two dimensional **planes** of interacting atoms or molecules, or of three-dimensional **networks** of interacting atoms or molecules, undergoing a cooperative order disorder or displacive processes. In addition to the dimensionality of the system one must take into consideration the dimensionality of the elementary units undergoing the phase transition: one may have one-dimensional (Ising), two dimensional (X-

Y) or three-dimensional (Heisenberg) cooperative arrangements of the unit atoms or molecules themselves. Fluctuations may play a major role in the order-disorder processes taking place close to the transition temperature T_C. The universality class of a phase transition is determined by the dimensionality of the system and the dimensionality of the order parameter associated to the elementary components of the system.

At the phase transition anomalies in a number of physical properties of the system become apparent: In particular, anomalies in the mechanical properties, the thermal properties (specific heat, thermal conductivity), magnetic properties, electrical properties (dielectric permittivity, conductivity, optical properties), etc. The structure and the time dependence of the physical properties of the system undergoing a phase transition can change remarkably at $T \cong T_C$.

***

The explosive growth in computer power which has taken place in the last few decades, and which is still taking place at the beginning of the 21th century, has found spectacular applications in Condensed Matter Physics.

As pointed out by Kurt Binder and Dieter W. Heermann in "Monte Carlo Simulation in Statistical Physics: An Introduction", the increasing application of numerical methods in Materials Physics and Condensed Matter Physics by researchers in academic centers and in industry are paying off very well.

In the following chapters (2.2 to 2.5) numerical results for Ising lattices of various dimensionalities obtained by means of careful analyses of Monte Carlo data for systems of 10^6 spins taken at close temperature and field intervals in the close vicinity of the transition point are shown to lead to excellent results, much better than those obtained by other methods. In particular, Onsager's calculated values for the critical exponents of two dimensional Ising lattices are reproduced exactly, and fractional critical exponents for three dimensional lattices are obtained, which fulfil very well the required scaling relationships.

Previous published work on **ferroelectric dipole waves** unrelated to Monte Carlo calculations, fully analogous to elementary excitations in ferromagnets, "magnons" are reported in chapter 2.6.

In "On the Critical Isotherms in Ising Systems" in *Ferroelectrics* **426** (2012) pp 166 published under the title "Monte Carlo Study of Critical Isotherms" by J.G. García, M.I. Marqués and J.A. Gonzalo, data on magnetization versus field exactly at T_C are shown to provide direct accurate results on the critical exponent δ^{-1} giving magnetization as a function of field for one, two, three and, four dimensional lattices, resulting in $\delta_{1D}^{-1} = 0$, $\delta_{2D}^{-1} = 0.0666 \cong 1/15$, $\delta_{3D}^{-1} = 0.1997(4) \cong 1/5$, $\delta_{4D}^{-1} = 0.332(5) \cong 1/3$.

In "On the 3D Critical Exponents for Ising Systems", previously published under the tile "Accurate Monte Carlo Critical exponent for Ising lattices", *Physica A*, **326**, 464 (2003) by Jorge García and Julio A. Gonzalo, it is shown that the critical exponents for three dimensional Ising lattices are given by $\beta_{3D} = 0.3126(4) \cong 5/16$, $\delta_{3D}^{-1} = 0.1997(4) \cong 1/5$ and $\gamma_{3D} = \beta_{3D}(\delta_{3D} - 1)$, the latest exponent confirming the expectation of the simple fraction for $\gamma_{3D} \cong 5/4$ done by Domb very early, and then discarded for no conclusive reason.

It is shown that the fractional values obtained for the full set of critical exponents fulfil the expected scaling relationships.

In "On the Equation of State for 3D Ising Systems", previously published work in *Ferroelectrics* **314**, 1-6 (2005) by J. García, M.I. Marqués and J. A. Gonzalo, accurate Monte Carlo data from a set of close isotherms near the critical point are properly analyzed.

It is shown that plots of data of $ML^{\beta\delta}$ vs $|t|L^{1/\nu}$ for increasing L values (L = 30, 60, 90, 115) scale very accurately and fall on top of each other, which is tantamount to stablishing the **equation of state**. High quality experimental data on ferromagnetic $CrBr_3$ by Ho & Litster are also examined.

In "Effective Critical Exponents at Minimal Dimensionalities" work not previously published by Mª Felisa Martínez, Carlos García and Julio A. Gonzalo on Ising stripes $D \times L$ ($D \ll L$) is given in 2.5.1, 2.5.2 and 2.5.3.

Finally, in "On Ferroelectric Dipole Waves", previously published under the tile "Ferroelectric elementary excitations at low temperatures: dipole waves", J. A. Gonzalo, *J. Phys. C: Solid State Phys* 20, **3985** (1987) Copyright © 1987 / OP Publishing Ltd, it is shown that at very low temperatures, much lower than the ferroelectric Curie temperature ordinary acoustic phonons can coexist with low energy spin-wave-like excitations in solids which possess an spontaneous polarization. These excitations are shown to contribute near 0 °K a specific heat term proportional to $T^{3/2}$, analogous to that characteristic of ferromagnetic magnons.

Lars Onsager (1903 – 1976)

CHAPTER 2.2
ON THE CRITICAL ISOTHERMS IN ISING SYSTEMS.

Monte Carlo investigations of magnetization versus field, $M_c(H)$, at the critical temperature provide direct accurate results on the critical exponent δ^{-1} for one, two, three and four-dimensional lattices:

$\delta_{1D}^{-1} = 0$, $\delta_{2D}^{-1} = 0.0666(2) \simeq 1/15$, $\delta_{3D}^{-1} = 0.1997(4) \simeq 1/5$, $\delta_{4D}^{-1} = 0.332(5) \simeq 1/3$.

This type of Monte Carlo data on δ, which is not easily found in studies of Ising lattices in the current literature, as far as we know, defines extremely well the numerical value of this exponent within very stringent limits.

Introduction

The spin -1/2 Ising model[1] has been a remarkably succesful model of a short-range interacting system to study phase transitions in magnetic, order-disorder, ferroelectric systems, etc. In this model, the spin variable s_i is allowed to take values ±1 on each lattice site. The Hamiltonian that rules the interaction between the spins has the following form

$$\mathcal{H} = -J\sum_{<ij>} s_i s_j - -H\sum_i s_i \qquad (1)$$

The phase transition of the system results as a consequence of the cooperative behavior of the spins determined by the first term of this hamiltonian, where J is the exchange energy. The second term contains the magnetic field H, which also influences the ordering of the spins.

The critical exponents of the Ising universality class have been extensively investigated for a long time by various methods, High (HT) and Low Temperature (LT) expansions, Monte Carlo (MC) simulations, Field Theoretical (FT) methods, etc. In particular, near the critical point $T \rightarrow T_c$ and $H \rightarrow 0$, the susceptibility (χ) exponent ($\gamma = \partial \log \chi^{-1}/\partial \log |T - T_c|$), the correlation length (ξ) exponent ($\nu = \partial \log \xi^{-1}/\partial \log |T - T_c|$), the pair correlation function [$\Gamma(r)$] exponent η (where

$\Gamma(r) \sim 1/\mathrm{rd}^{-2+\eta}$), the specific heat exponent ($\alpha = \partial \log C^{-1}/\partial \log |T - T_\mathrm{c}|$), and the spontaneous magnetization exponent ($\beta = \partial \log M/\partial \log |T - T_\mathrm{c}|$) have been repeatedly investigated by MC methods, but no direct MC investigations of the critical isotherm exponent ($\delta = \partial \log H/\partial \log M$) have been published, as far as we know.

Monte Carlo Method

The determination of the critical exponent δ from Monte Carlo data at the critical isotherm, $M_\mathrm{c}(H)$ at $T = T_\mathrm{c}$, was made in such a way as to minimize (1) finite size effects, appearing only at H very close to zero, and saturation effects, appearing at the opposite end of the sets of data, i.e. at H away from zero, when the power law $M \sim H^{1/\delta}$ loses its validity. Relatively large lattices (of the order of 10^6 spins or more) with periodic boundary conditions and very closely spaced field intervals ($\Delta H \sim 0.0005$) were used to approach the critical point.

In order to confirm the value of the critical temperature for each dimensionality, $M_\mathrm{s}(T)$ was first determined by means of a Wolff[2] (modified Swendsen-Wang[3]) algorithm. However, $M_\mathrm{c}(H)$ was studied with a standard Metropolis[4] algorithm which produces an excellent local and global thermalization of the whole lattice for one single temperature (in our case always $T = T_\mathrm{c}$). In all cases, 100,000 Monte Carlo steps were used in order to insure good thermal equilibrium, and 20,000 states were considered in the partition function at each temperature/field.

It is easy to understand that our Metropolis algorithm, modified for studying the evolution of magnetization with magnetic field, is sufficient to investigate δ. When we make calculations modifying H by very small amounts one would not expect too large changes in the fluctuations, making it unnecessary to use modified clusters algorithms, which get statistically more independent states for the evolution with temperature, as it is well known. In any case, the good quality of the results shown below, obtained with the Metropolis algorithm, speak for themselves.

The $M_c(H)$ data allow the direct determination of δ^{-1} from log-log plots of M vs. H at $T = T_C$.

One-dimensional Ising critical isotherm

The one-dimensional case is the easiest of all to be considered because the critical temperature is exactly zero (there is no real phase transition). In Figure 1, we show that the smallest non-zero value of the magnetic field H make the whole lattice turn in the same direction, perfectly ordered, keeping the magnetization at the maximum value ($M = 1$) all the way.

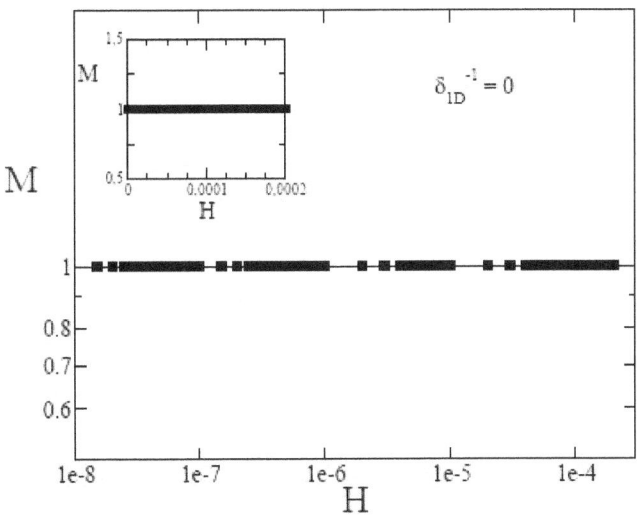

Fig. 1. Magnetization (M) vs. field (H) at $T = T_C = 0$ for a one-dimensional (d = 1)

Two-dimensional Ising critical isotherm

In 1944, Onsager[5] performed the calculation of the exact partition function of the two-dimensional Ising model in zero field. However, the same problem in a magnetic field remain unsolved although we can study their behavior numerically. In Figure 2, numerical Monte Carlo $M_C(H)$ data allow the direct determination of $\delta_{2D}^{-1} = 0.0666(2)$ from log-log plots of M vs. H at $T = T_C = 2.269185314213$.

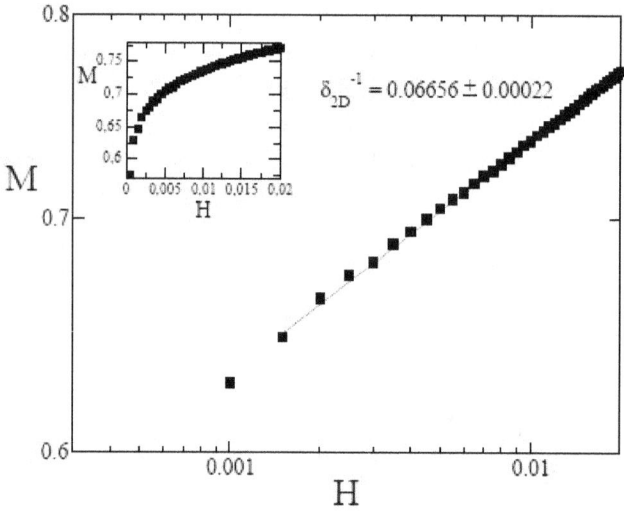

Fig. 2. Magnetization (M) vs. field (H) at $T = T_C = 2.269185314213$ for a two-dimensional ($d = 2$) Ising squared monolayer. The slope gives the $\delta_{2D}^{-1} = 0.0666(2)$ value for the critical isotherm exponent.

Three-dimensional Ising critical isotherm

The Ising criticality for dimension $d = 3$ has been studied using several theoretical approaches (including MC simulations) that have resulted in an (almost complete) set of safe estimates of the critical exponents. Some exponents like γ or β have received special attention and their best estimates have been summarized[6] as $\beta_{3D} \simeq 1.2372(5)$ and $\gamma_{3D} \simeq 0.3265(3)$. However, in the case of the critical exponent δ_{3D}, there is no direct MC measure of its value and the currently accepted estimate have been determined using the scaling relation $\gamma = \beta(\delta-1)$ resulting in a value of $\delta_{3D} = 4.789(2)$.

As it is shown if Figure 3, our Monte Carlo Mc(H) data allow the determination of $\delta_{3D}^{-1} = 0.1997(4)$ from log-log plots of M vs. H at the currently accepted value[7] of the critical temperature $T = T_C = 4.511523785$. Note that this direct value of the critical exponent $\delta_{3D} \simeq 5$ is appreciably different from the actual[6] estimated value 4.789(2).

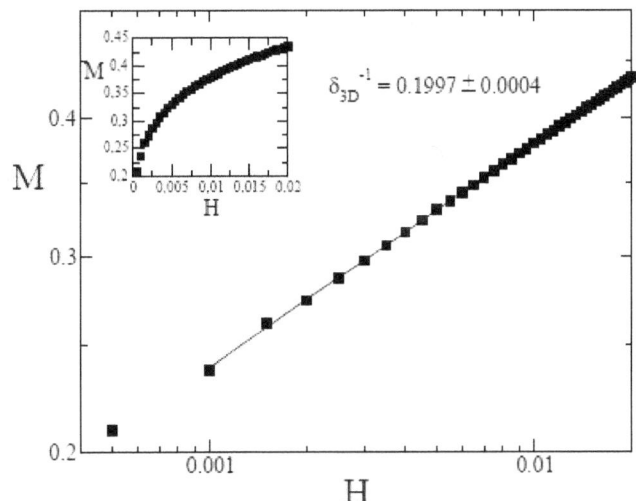

Fig. 3. Magnetization (M) vs. field (H) at $T = T_C = 4.511523785$ for a three-dimensional ($d = 3$) Ising cube. The slope gives the $\delta_{3D}^{-1} = 0.1997(4)$ value for the critical exponent.

Four-dimensional Ising critical isotherm

It is important to note at this point, that for dimensions $d \geq 4$ the exponents of the Ising model become the same and take the corresponding mean-field values[8]. These exponents suddenly lock into a set of values that become independent of the dimensionality.

In Figure 4, our Monte Carlo Mc(H) data allow the determination of $\delta_{4D}^{-1} = 0.332(5)$ from log-log plots of M vs. H at the currently accepted value[9] of the critical temperature $T = T_C = 6.68029(9)$. It must be noted that finite size effects begin to be observable just at $H \leq 0.001$ for the particular lattice size used.

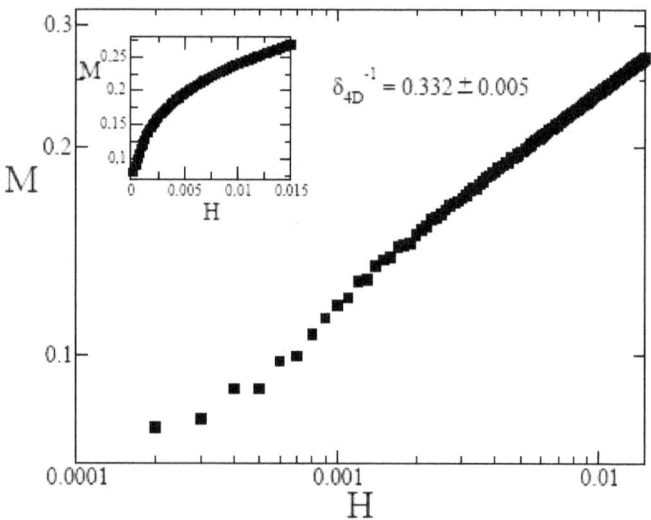

Fig. 4. Magnetization (M) vs. field (H) at $T = T_C = 6.68029$ for a four-dimensional ($d = 4$) Ising hypercube. The slope gives the $\delta_{4D}^{-1} = 0.332(5)$ value for the critical exponent.

Concluding remarks

Table I summarizes our results for Ising systems ($1 \leq d \leq 4$) showing excellent agreement between our MC results for δ and the respective integer values $\delta(d)$ indicated.

d	1	2	3	4
T_c	0	2.269185314213	4.511523785	6.68029
δ_{MC}^{-1}	0	0.0666(2)	0.1997(4)	0.332(5)
$\delta(d)$	∞	15	5*	3

Table I. Numerical values for transition temperature (T_C) and critical isotherm exponent (δ_{MC}^{-1}) for Ising systems with $1 \leq d \leq 4$. *Note that the value for $d = 3$ is appreciably different from the currently accepted value $\delta_{3D} = 4.789(2)$[6].

Acknowledgments. – We specially acknowledge helpful comments and software by M.I. Marqu´es. Support from the Spanish DGICyT through Grant Number BFM2000-0032 is grate- fully acknowledged.

REFERENCES

[1] J.M. Yeomans, *Statistical Mechanics of Phase Transitions* (Oxford University Press) 1992.
[2] U. Wolff, Phys. Rev. Lett., 62 (1989) 361.
[3] R.H. Swendsen and J.S. Wang, Phys. Rev. Lett., 58 (1987) 86.
[4] N. Metropolis, A.W. Rosenbluth, M.N. Rosenbluth, A.H. Teller and E. Teller, J. Chem. Phys., 21 (1953) 1087.
[5] L. Onsager, Phys. Rev., 65 (1944) 117.
[6] A. Pelissetto and E. Vicari, Phys. Rep., 368 (2002) 549-727.
[7] H.W.J. Blöte, L.N. Shchur and A.L. Talapov, Int. J. Mod. Phys. C, 10 (1999) 137.
[8] J.A. Gonzalo, *Effective Field Approach to Phase Transitions and some Applications to Ferroelectrics* (World Scientific, Singapore) 1991.
[9] D. Stauffer and J. Adler, Int. J. of Mod. Phys. C, Vol. 8, 2 (1997) 263-267.

CHAPTER 2.3
ON THE 3D ISING CRITICAL EXPONENTS.

A careful Monte Carlo investigations of the phase transition very close to the critical point ($T \to T_C$, $H \to 0$) in relatively large $d = 3$, $s = 1/2$ Ising lattices did produce critical exponents $\beta_{3D} = 0.3126(4) \approx 5/16$, $\delta_{3D}^{-1} = 0.1997(4) \approx 1/5$ and $\gamma_{3D} = 1{:}253(4) \approx 5/4$. Our results indicate that, within experimental error, they are given by simple fractions corresponding to the linear interpolations between the respective two dimensional (Onsager) and four dimensional (mean field) critical exponents. An analysis of our inverse susceptibility data $\chi^{-1}(T)$ vs. $|T-T_C|$ shows that these data lead to a value of γ_{3D} compatible with $\gamma' = \gamma$ and $T_C = 4.51152(12)$, while γ values obtained recently by high and low temperature series expansions and renormalization group methods are not.

Introduction

The Ising model, see, e.g. Ref[1], rightly considered as the prototype of statistical systems with non-classic power law critical behavior, has been extensively investigated for many years. Systems with short-range interactions display Ising like critical behavior, e.g. liquid-vapor, multicomponent mixtures, uniaxial magnets, etc., and there is a wealth of very accurate experimental information on these systems.

To describe e.g. the behavior of a $s = 1/2$ uniaxial ferromagnetic near the critical point ($T = T_C$, $H = 0$) two critical exponents, β for the spontaneous magnetization $M_S(T)$ and δ^{-1} for the field dependence of the magnetization $M_C(H)$ at the critical temperature, determine basically the critical behavior through $M_S(T) \sim |T_C - T|^\beta$ and $M_C(H) \sim H^{1/\delta}$ It is well known[2] that for a two dimensional Ising lattice ($d = 2$) Onsager's solution gives fractional values for $\beta_{2D} = 1/8$ and $\delta_{2D}^{-1} = 1/5$. For a four dimensional Ising lattice ($d = 4$) on the other hand, the critical exponents are the mean field exponents[3], given also by fractional values, $\beta_{4D} = 1/2$ and $\delta_{4D}^{-1} = 1/3$.

It is a legitimate question to ask whether for a three dimensional Ising lattice ($d = 3$), for which no general theoretical solution is available for the moment, the values for β_{3D} and δ_{3D}^{-1} are rational fractions or not. In fact, almost 40 years ago, Cyril Domb, one of the very pioneers in the then rapidly growing field of phase transitions, suggested that for three dimensional Ising lattices the susceptibility critical exponent $\gamma = \beta(\delta - 1)$ might be given by the fractional value $\gamma_{3D} = 5/4 = 1.25$. Since then a tremendous amount of work (experimental, theoretical and computational) has been performed with the aim to get ever more precise numerical characterizations of the phase transitions. Table 1 gives a representative sample, see e.g. Ref.[4], of numerical values[5-12] for the exponents γ_{3D} and β_{3D} obtained by various methods: high temperature expansion series, low temperature expansion series, Monte Carlo simulations and field theoretical methods. The overall picture of the numerical values for γ_{3D} and β_{3D} is reasonably good, and they seem to favor ftp γ_{3D} < 1.25 and β_{3D} > 0.3125, but, clearly, the uncertainties quoted in parenthesis cannot be taken strictly at face value.

In the present work we present results on critical values based upon optimized accurate Monte Carlo calculations and we investigate to what extent the $d = 3$ Ising exponents are compatible with the simple fractions interpolated between the fractional $d = 2$ Ising exponents and the, fractional too, $d = 4$ Ising exponents. In particular we will use direct determinations of β_{3D} from $M_S(T)$ data at $H = 0$, and of δ_{3D}^{-1} from $M_C(H)$ data at $T = T_C$, as well as, $\chi^{-1}(T)$ vs including data both below T_C (LT phase) and above T_C (HT phase), which allow us to make internal consistency checks, so that the actual value used for the critical temperature can be confirmed to be compatible with the scaling requirement $\gamma' = \gamma$ or not.

Table 1. Some estimated $s = 1/2$ critical exponents for $d = 3$ Ising simple lattices

Method	Ref.	Year	γ	β
High T expansions	[5]	(2002)	1.2368(10)	0.3243(30)[a]
Low T expansions	[6]	(1997)	1.2388(10)	0.3278(13)[a]
Low T expansions	[7]	(1993)	1.251(28)	0.329(9)
Monte Carlo	[8]	(1991)	1.255(10)	0.320(3)
Monte Carlo	[9]	(1999)	1.2372(13)[b]	0.3269(5)
Field theoretical	[10]	(2000)	1.255(18)[b]	0.325(5)
Field theoretical	[11]	(2001)	1.2403(8)	0.3257(5)
Field theoretical	[12]	(1994)	1.258	0.336

[a]Obtained using the scaling relation $\beta = \nu(1+\eta)/2$
[b]Obtained using the scaling relation $\gamma = (2-\eta)/\nu$

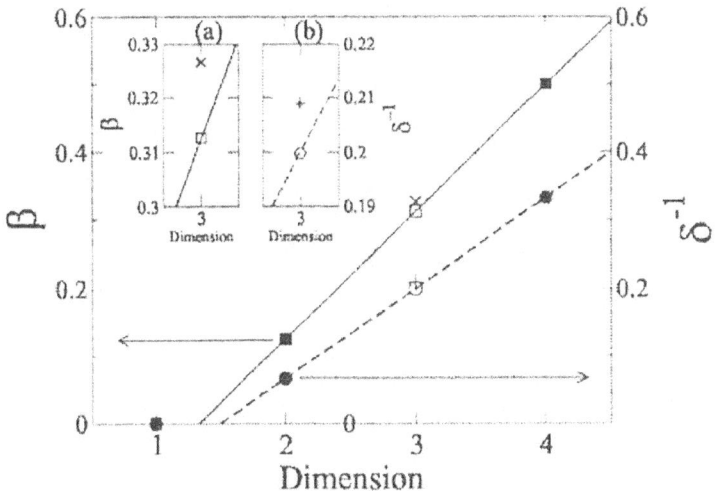

Fig. 1. β_{3D} and δ_{3D}^{-1} (open symbols) obtained by interpolation between exact two-dimensional data from Onsager[2] and four-dimensional data from mean field

theory, see e.g. Ref.[17], (filled symbols). The crosses show current accepted results[5]. In all cases, error bars are smaller than the symbol's size. Insets (a) and (b) are a blow-up close to the $d = 3$ region for β and δ^{-1}, respectively.

In order to establish the reliability of the data and the propriety of the method of analysis used we will proceed in two steps. First we will check data on large two dimensional lattices, for which the fractional values of the exponents are known exactly, and then we will analyze data on large three dimensional lattices for which the fractional values proposed arc only educated guesses. Our data, therefore, can either lend support or leave unsupported the fractional values proposed. Fig. 1. shows the evolution of critical exponents with lattice dimensionality $d = 2, 3, 4,...$ for Ising systems.

Monte Carlo determinations of $M_S(T)$ and $M_C(H)$

Finite size scaling Monte Carlo simulations of phase transitions are known to be among the best and more effective techniques available to characterize the critical behavior of model systems such as Ising systems of any dimensionality. To determine critical exponents with the highest possible reliability it is desirable (i) to increase systematically the size L of the system till the results become practically independent of L within error bars; (ii) to make the interval (temperature or magnetic field) between successive states in the vicinity of the critical point as small as possible, choosing wisely the full range (temperature of magnetic field) so that is not too large (time consuming) or too small (inconvenient of incomplete) to determine adequately the exponent in question; (iii) to insure that the time spent in the calculation at each point (temperature, magnetic field) is sufficient to arrive to true equilibrium. We did use two dimensional lattices of 800^2, 900^2 and 1000^2 spins, and three dimensional lattices of 90^3 100^3 and 115^3 spins.

The temperature intervals were of the order of 0.001, in

units of T such that T_C^{2D} = 2.269185314213... and T_C^{3D} = 4.51152(12) for determinations of the spontaneous magnetization $M_S(T)$, and field intervals of the order of 0.0005 for determinations of $M_C(H)$ at the critical isotherm.

The number of Monte Carlo steps taken to insure equilibrium at each state was 50,000 and the number of states considered in the partition function was 10,000. No improvement was detected by increasing the number of states. To get $M_S(T)$ a Wolff[13] algorithm was used, which is a generalization of the Swendsen-Wang[14] algorithm. To obtain $M_C(H)$, a somewhat novel but straightforward calculation not usually found in the literature, a standard Metropolis[15] algorithm was used and it was proved to be quite adequate.

Initially both periodic boundary conditions and free boundary conditions were used. After confirming that for very large lattices the difference between results obtained with either set of boundary conditions was quite negligible, we used subsequently periodic boundary conditions all the way.

For the temperature scans we always began at $T > T_C$ and spent sufficient time at the beginning to make sure that thermal equilibrium was attained already far above the transition temperature. Nevertheless the statistical ups and downs in the residual $M(T)$ data at $T > T_C$, are much larger than the corresponding fluctuations at $T < T_C$ which become almost negligible and display a beautiful continuity all the way down in temperature. The log-log plots of M vs. $|T - T_C|$ and M vs H indicate clearly, first that we have been able to get really close to the critical point, more so than in other Monte Carlo calculations for which data in the literature are sufficiently explicit, and, second, that in the full range displayed for T or H corrections to scaling are invisible, which make the data especially apt to determine the exponents β_{3D} and δ_{3D}^{-1}.

Monte Carlo results for two dimensional lattices

Fig. 2(a) gives a log-log plot of the spontaneous magnetization as a function of temperature for a squared Ising lattice with 1000 x 1000 spins. Finite size effects show up as rounding at $|T - T_C| \to 0$ which occurs only at $|T - T_C| \leq 0{:}002$ with $T_C = 2.269185314213...$. Corrections to scaling should appear at the other end of the temperature range examined, but are quite invisible in our data. The spontaneous magnetization data give directly a value for $\beta_{2D} = 0.1242 \pm 0{:}0008 \approx 1/8$, as expected. Fig. 2(b) presents a log-log plot of the magnetization as a function of field for the same squared Ising lattice with 1000 x 1000 spins at the critical isotherm. Finite size effects begin to appear as incipient rounding at $H \leq 0.0001$ but are almost imperceptible. Corrections to scaling should appear at the other end of the field range investigated but are completely negligible in our data. These critical isotherm data, not previously investigated in depth, as far as we know, give directly $\delta_{2D}^{-1} = 0.06656 \pm 0.00022 \approx 1/15$ as expected.

These results lend support to the expectation that carefully taken Monte Carlo data in large enough lattices taken at small enough temperature/field intervals are accurate enough to investigate whether critical exponents are given by simple fractions or not, at least in the case of two dimensions.

A straightforward numerical analysis of the fractions compatible with the experimental results and the uncertainties quoted has been made. For $d = 2$ it can be seen that n/m

with $n < m$ compatible with the uncertainties, must go to m values very large, $m > 256$ for β_{2D} and $m > 360$ for δ_{2D}^{-1}

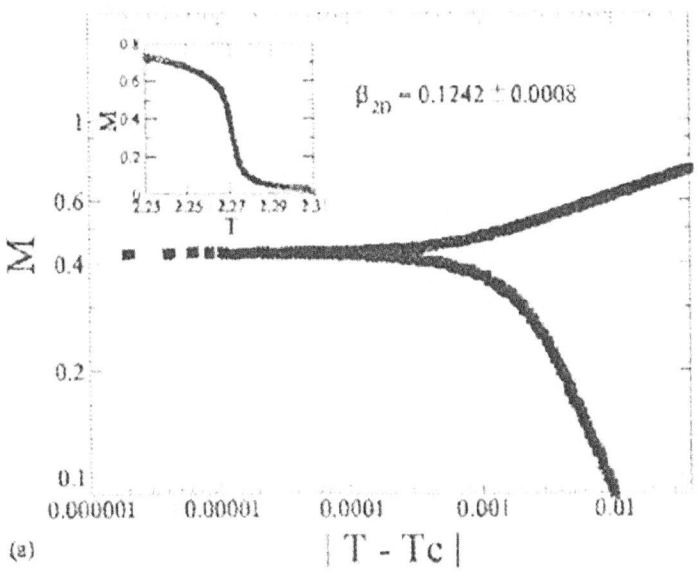

(a)

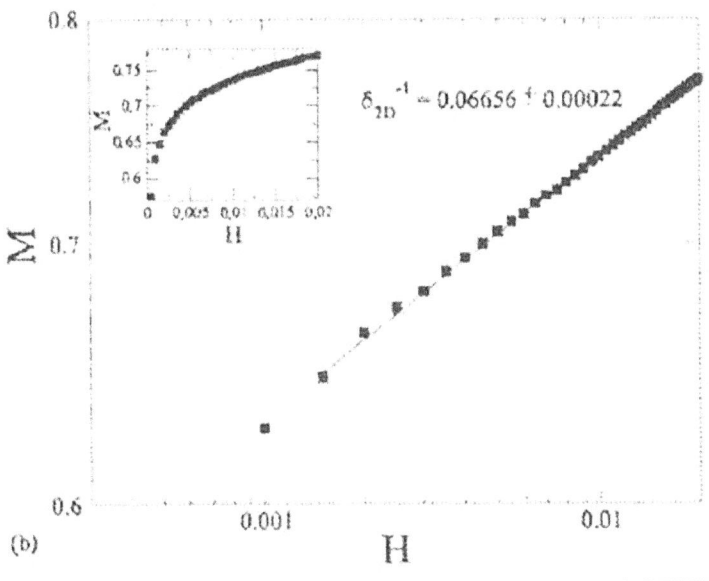

(b)

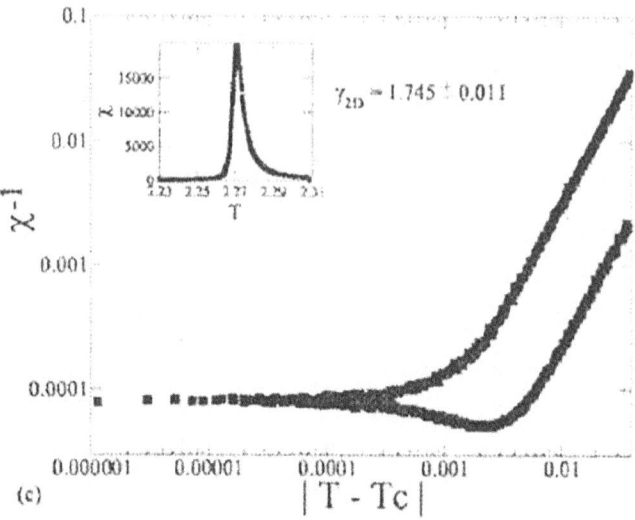

Fig. 2(a-c) Two dimensional critical exponents obtained by fits of Monte Carlo data of spontaneous magnetization $M_S(H=0)$ as a function of temperature T near T_C = 2.269185314213, critical isotherm, M as a function of H at $T = T_C$, and isothermal susceptibility χ, respectively, in a system of 1000 x 1000 spins. Insets show the row data.

Monte Carlo results for three dimensional lattices

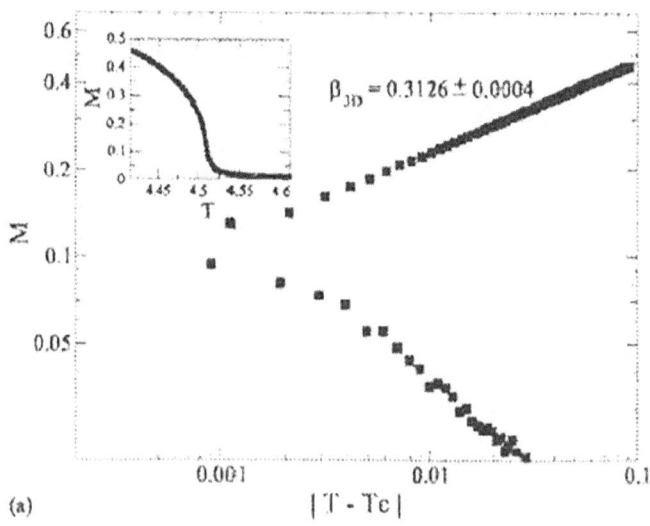

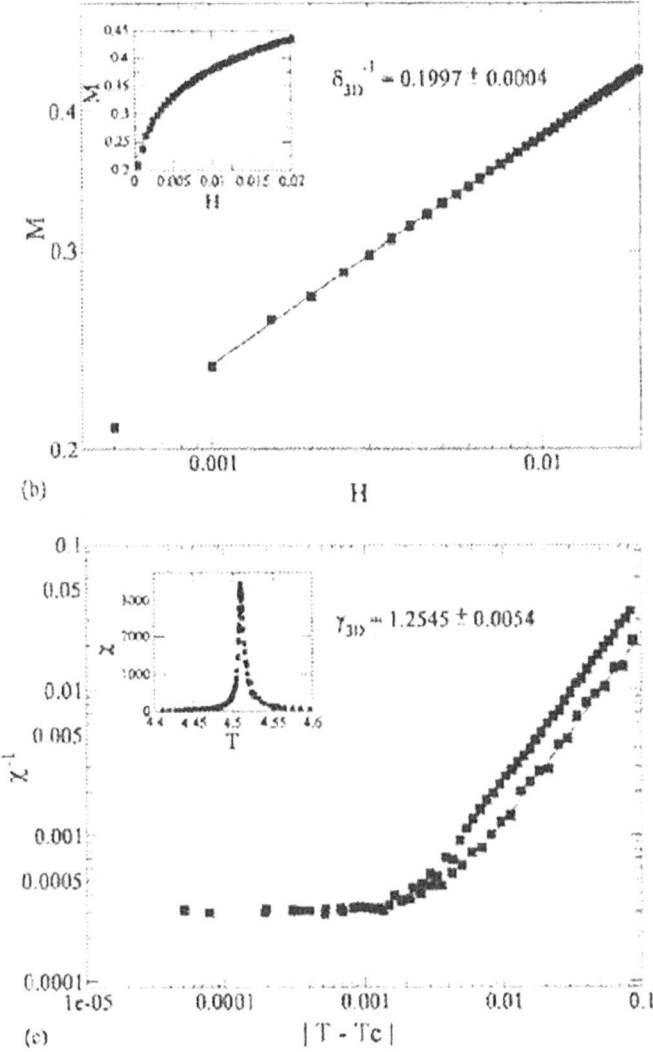

Fig. 3. Three dimensional critical exponents $\gamma_{3D}(a)$ $\beta_{3D}(b)$ and $\delta_{3D}^{-1}(c)$ obtained by fits of Monte Carlo data of spontaneous magnetization M_S near T_C = 4.51152 ± 0.00012, critical isotherm and isothermal susceptibility χ, respectively, in a system of L^3 = 115 × 115 × 115 spins with periodic boundary conditions. Insets show the row data.

Fig. 3(a) gives a log-plot of $M_S(T)$ vs. T for a simple cubic lattice with 115 x 115 x 115 spins. Lattices with 90 x 90 x 90 and 100 x 100 x 100 spins were investigated and the results, of course slightly less accurate, where completely consistent with those obtained with the larger lattice. Again finite size effect appear at, and they appear clearly at $|T - T_C| \leq 0.004$ but corrections to scaling are invisible in the temperature range investigated. For β_{3D} we get

$$\beta_{3D} = 0.3126(4) \cong \frac{5}{16} \tag{1}$$

Fig. 3(b) produces a similar log-log plot of the magnetization as a function of field for the same lattice with $115 \times 115 \times 115$ spins at $T = T_C = 4.51152(12)$. It is clear that both finite size effects and systematic departures from scaling are absent or imperceptible in the field range explored. These new, not previously directly investigated critical isotherm data, result in

$$\gamma_{3D}^{-1} = 0.1997 \cong \frac{1}{5} \tag{2}$$

which differs somewhat from previous indirect estimates δ_{3D} summarized by Pelissetto and Vicari[4] as giving $\delta_{3D} = 4.789(2)$, equivalent to $\delta_{3D}^{-1} = 0.2088(1)$.

Using the numerical values for and β_{3D} and δ_{3D}^{-1} in Eqs. (1) and (2) one gets γ_{3D} indirectly as

$$\gamma_{3D} = \beta_{3D}(\delta_{3D} - 1) = 1.252 \tag{3}$$

by means of the scaling relation for γ in terms of β and δ.

But one can get γ directly from plots of susceptibility data at $T > T_C(\gamma)$ as well as at $T > T_C(\gamma')$ giving us the opportunity to check that the fundamental equality $\gamma = \gamma'$, confirmed by renormalization group theory[4], is fulfilled.

Fig. 3(c) gives log-log plots of direct Monte Carlo data on the susceptibility $\chi^{-1}(T)$ as a function of $|T - T_C|$ both above and below the critical temperature $T_C = 4{:}51152$. The resulting directly determined value of gamma is

$$\gamma_{3D} = \gamma'_{3D} = 1.252 \tag{4}$$

which is in excellent agreement with the forty-year-old prediction of Domb and Sykes[16].

As mentioned before the following consistency check was made: changing smoothly the critical temperature value by increments (decrements) of 0.00002 we can obtain smooth changes in the effective values of the exponents γ_{3D} and γ'_{3D} to reproduce numerical values as low as 1.237 for, resulting $\gamma'_{3D} > 1.25$, or as high as 1.263 for, resulting in $\gamma'_{3D} < 1.25$.

With our Monte Carlo data, which in sufficiently wide ranges appear to be free of finite size effects as well as of corrections to scaling effects, only using the right $T_C = 4.51152$, in perfect agreement with the values quoted in the literature, the requirement $\gamma = \gamma'$ is duly fulfilled.

Table 2 shows that according to our data for simple cubic $s = 1/2$ Ising lattices, simple fractions for β and δ^1 (the same procedure could be employed to check simple fractions for) compatible with the quoted uncertainties are given always by n/m fractions which are, either identical to the interpolated values $\beta_{3D} = 5/16$, $\delta_{3D}^{-1} = 1/5$ ($\gamma_{3D} = 5/4$) or much more complex fractions involving $m > 128$ (for β) and $m > 285$ (for δ^1).

Table 2. Monte Carlo $s = 1/2$ Ising critical exponents in a three dimensional simple cubic lattice of 115x115x115 spins

β	δ^{-1}
0.31254 ± 0.00029	0.1997 ±0.0004

Finite size scaling analysis

Fig. 4 shows a scaling plot of Ising 3D Monte Carlo data for $L = 90, 100, 115$. Here ε is defined as $|(T - T_C)/T_C|$, the renormalized temperature. It may be seen that a surprisingly good scaling is achieved with the fractional exponents $\beta_{3D} = 5/16$ and $v_{3D} = 5/8$. As far as we know, no such quality scaling plots obtained with Monte Carlo data are common in the current literature for Ising three-dimensional systems.

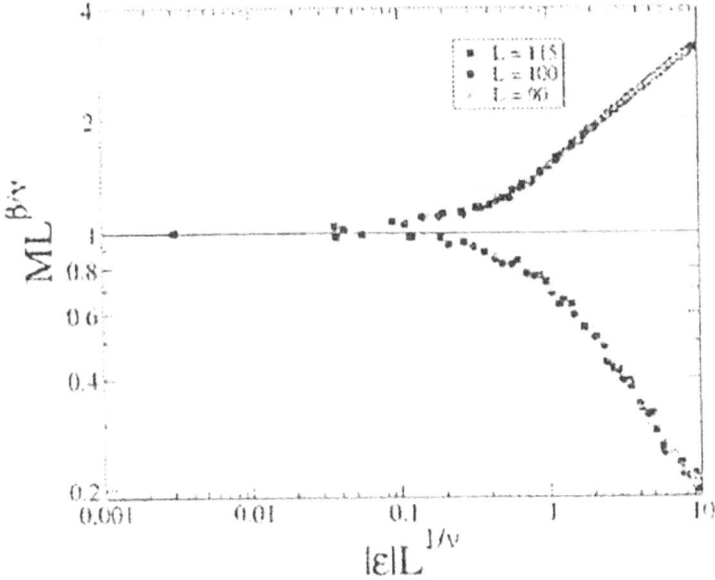

Fig. 4. Scaled magnetization $ML^{\beta/\nu}$ vs. scaled temperature $|(T-T_c)/T_c|L^{1/\nu}$ for three dimensional Ising lattices with $L^3 = 90^3$, 100^3 and 115^3 spins ($\beta_{3D} = 5/16$, $\nu_{3D} = 5/8$, $T_c = 4.51152(12)$).

Concluding Remarks

Our Monte Carlo data do not prove directly beyond doubt that three dimensional Ising lattices are characterized by the fractional critical exponents $\beta = 5/16$ and $\delta_{3D}^{-1} = 1/5$ (and through the corresponding scaling relations, by the resulting critical exponents $\gamma = 5/4$, $\alpha = 1/8$, $\nu = 5/8$ and $\eta = 0$) but they support strongly and consistently this fractional values in particular the susceptibility data resulting in $\gamma_{3D} = \gamma'_{3D}$ $=1.25 = 5/4$. Our empirical data may be the basis for future well grounded theoretical arguments confirming the above fractional critical exponents.

Acknowledgements: We acknowledge helpful comments by Lidia Braunstein, Gerry Paul and Manuel 1. Marques to Jorge Garcia during fruitful stays at the Boston University Physics Department. Support from the Spanish Ministry of Science and Technology through Grant Number BFM2000-0032 is gratefully acknowledged.

REFERENCES

[1] J. Adler, J. Phys. A 16 (1983) 3585.
[2] L. Onsager, Phys. Rev. 65 (1944) 117.
[3] J.M. Yeomans, Statistical Mechanics of Phase Transitions, Oxford University Press, Oxford, 1992.
[4] A. Pelissetto, E. Vicari, Phys. Rep. 368 (2002) 549.
[5] P. Butera, M. Comi, Phys. Rev. B 65 (2002) 144 431.
[6] P. Butera, M. Comi, Phys. Rev. B 56 (1997) 8212.
[7] A.J. Guttmann, I.G. Enting, J. Phys. A 26 (1993) 806.
[8] J. Oitman, C.J. Hamer, W. Zheng, J. Phys. A 24 (1991) 2863.
[9] H.W.J. Blote, L.N. Shchur, A.L. Talapov, Int. J. Mod. Phys. C 10 (1999) 137.
[10] N. Ito, K. Hukushima, K. Ogawa, Y. Ozeki, J. Phys. Soc. Japan 69 (2000) 1931.
[11] F. Jasch, H. Klcinert, J. Math. Phys. 42 (2001) 52.
[12] J. Berges, N. Tetradis, C. Wetterich, Phys. Rev. Lett. 77 (1996) 873.
[13] U. Wol, Phys. Rev. Lett. 62 (1989) 361.
[14] R.H. Swendsen, J.S. Wang, Phys. Rev. Lett. 58 (1987) 86.
[15] N. Metropolis, A.W. Rosenbluth, M.N. Roscnbluth, A.11, 'feller, H. Teller, J. Chem. Phys. 21 (1953) 1087.
[16] C. Domb, M.F. Sykes, Phys. Rev. 128 (1962) 168 173.
[17] J.A. Gonzalo, Effective Field Approach to Phase Transitions and Some Applications to Ferroelectrics, World Scientific, Singapore, 1991.

Ernst Ising (1900 – 1998)

CHAPTER 2.4
ON THE EQUATION OF STATE FOR 3D ISING SYSTEMS.

Accurate Monte Carlo data from a set of isotherms near the critical point are analyzed using two *RG* based complementary representations given respectively in terms of $\bar{h} = h/|t|^{\beta\delta}$ and $\bar{\tau} = t/h^{1/\beta\delta}$. Scaling plots for data on simple cubic Ising lattices are compared with plots of $ML^{\beta\delta}$ vs $|t|L^{1/\nu}$ for increasing L values and with high quality experimental data on CrBr$_3$. Finite size effects and the equation of state are discussed.

The equation of state of a system undergoing a phase transition in the vicinity of the critical point is asymptotically given by the corresponding scaling function. Since the homogeneity assumption was formulated for the free energy of cooperative systems undergoing a phase transition efforts were made to cast properly the scaling function. The renormalization group approach provides the unifying picture for cooperative phenomena at phase transitions, but does not give direct hints regarding scaling functions and has not played a prominent role in the search for constructing explicit scaling functions. Monte Carlo methods[1] using Wolff algorithms[2] have been extensively used to describe $M(T)$ for $H=0$ in the whole range of temperatures both below and above the critical temperature (specially at $T \cong T_C$ but few, if any, Monte Carlo simulations of magnetic isotherms $M(H)$ at $T \cong T_C$, have been reported in the literature. It is clear, however, that with recent improvements in computing facilities (larger memory, greater speed, and better availability) accurate, well thermalized, closely spaced data can provide very substantial contributions to describing the phase transition and to better understanding of finite size effects.

Simulations of this type are reported in this work performed using Metropolis algorithms[3], which are specially convenient to describe the system evolution at constant temperature and for small field increments[4,5]. Using relatively large lattices with 70x70x70 spins, an accurate characterization of the scaling behavior, and, therefore, the equation of state in the vicinity of the critical point ($H = 0$, $T_C =$

4.511523785)⁶ for simple cubic Ising lattices can be obtained. Details of the Monte Carlo calculations for isotherms taken at T near T_C are given in reference [4]. Periodic boundary conditions were used and 140,000 Monte Carlo steps were taken at each field/temperature to ensure equilibrium.

For the scaling representation of the raw $M(H)$ data at each T we did use two complementary ways: (a) the usual way[7], involving $\bar{m} \equiv M(t,h)/|t|^{\beta\delta}$ and $\bar{h} = h/|t|^{\beta\delta}$ where $t \equiv (T-T_C)/T_C$ involves the temperature gap to T_C, h (reduced field) is proportional to H, β is the spontaneous magnetization critical exponent $\beta \equiv \partial \log M / \partial \log |t|$ at $T \to T_C$, and δ is the critical isotherm exponent $\delta^{-1} \equiv \partial \log M / \partial \log h$ at $T \to T_C$, and (b) a complementary way involving $\bar{\mu} \equiv M(t,h)/h^{1/\delta}$ and $\bar{\tau} = t/h^{1/\beta\delta}$.

Near the fixed point corresponding to the critical point, the singular part of the reduced free energy per spin may be written[7] in scaling form as

$$\bar{f}_s(g_1, g_2, g_3, \ldots) \sim \bar{b}^d \bar{f}_s(b^{y_1} g_1, b^{y_2} g_2, b^{y_3} g_3, \ldots) \quad (1)$$

where b is an arbitrary scale factor and the y_i relate to the usual critical exponents. To get (a) Eq. (1) is differentiated with respect to the field in the usual way to obtain

$$M(t, h, g_3, \ldots) \sim \bar{b}^{-d+y_2} M(b^{y_1} t, b^{y_2} h, b^{y_3} g_3, \ldots) \quad (2)$$

Taking $b^{y_1} |t| = 1$ and setting the irrelevant variables equal to zero

$$M(t,h) \sim |t|^{(d-y_2)/y_1} M\left(\pm 1, h|t|^{-y_2/y_1}\right) \quad (3)$$

and using the well known scaling relationships[7] giving y_1 and y_2 in terms of β and δ one gets

$$M(t,h) \sim |t|^{\beta} M\left(\pm 1, h|t|^{-\beta\delta_1}\right) \quad (4)$$

which implies scaling using the ordinary scaling variables

$$m \equiv M(t,h) \sim |t|^{\beta}, \quad \bar{h} \equiv h/|t|^{\beta\delta} \quad (5)$$

To get (b), on the other hand, we can argue in an analogous way. Taking $b^{y_2}h = 1$ in Eq. (2) and setting the irrelevant variables equal to zero,

$$M(t,h) \sim |t|^{(d-y_2)/y_2} M\left(t \cdot h^{-y_1/y_2}, 1\right) \quad (6)$$

and using y_1 and y_2 in terms of β and δ, one finally gets

$$M(t,h) \sim h^{1/\delta} M\left(t^{-1/\beta\delta}, 1\right) \quad (7)$$

which implies as alternative scaling variables

$$\bar{\mu} \equiv M(t,h)/h^{1/\delta}, \quad \bar{\tau} \equiv t/h^{1/\beta\delta} \quad (8)$$

Fig.1 (a) gives our Monte Carlo data using m and h as scaling variables as given by Eq. (5) with $T_C = 4.511523785$ and the exponents $\beta = 5/16 = 0.3125$ and $\delta = 5$. The usual behavior is observed. We may note that the data scale extremely well and that only very minor deviations at lower h attributable to finite size effects are perceptible. Fig.1(b) gives the same data using the alternative scaling representation. It may be noted immediately that the plot in Fig. 1 (b) resembles closely[1] typical scaling plots of $ML^{\beta/\nu}$ vs $|t|L^{1/\nu}$ for lattices with linear dimension L. in our case, for Ising systems of L^3 spins, suggesting[8] formal relationships between H and L, M and L, and $|t|$ and L, which, through the scaling plot branches defining the critical isotherm, the spontaneous magnetization (coexistence curve) below T_C, and the low field susceptibility above, imply respectively

$$H \sim L^{-\beta\delta/\nu}, \quad M \sim L^{\beta/\nu} \quad \text{and} \quad |t| \sim L^{1/\nu} \tag{9}$$

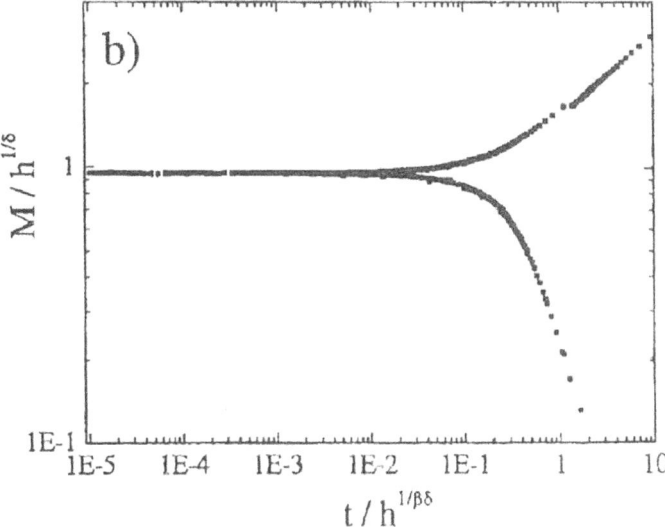

Fig.1 Scaling plots of Monte Carlo data for a s.c. Ising lattice of 70^3 spins. (a) $m \equiv M/|t|^\beta$ vs $\bar{h} \equiv h/|t|^{\beta\delta}$ and (b) $\bar{\mu} \equiv M/h^{1/\delta}$ vs $\bar{\tau} \equiv t/h^{1/\beta\delta}$. The data include 30 isotherms in the intervals $4 < T < 5$ and $0 < h < 0.02$, the same symbol has been used for all of them. The critical temperature[6] was taken as $T_C = 4.511523785$ and the critical exponents[4] $\beta = 5/16$ and $\delta = 5$.

Hence

$ML^{\beta/\nu} \sim \text{const} \to$ critical isotherm for $(t <> 0)$

$ML^{\beta/\nu} \sim \text{const}\left(|t|L^{1/\nu}\right)^{\beta} \to$ spontaneous magnetization for (t < 0)

$ML^{\beta/\nu} \sim \text{const}\left(|t|L^{1/\nu}\right)^{\beta(\delta-1)/2} \to$ susceptibility for (t > 0)

Fig.2 shows data of $ML^{\beta/\nu}$ vs $|t|L^{1/\nu}$ for $H = 0$ and $L = 30$, 60, 90, 115 which mimic the behavior shown in Fig. 1(b) implying that simulations using periodic boundary conditions of phase transitions with finite size show the effects of an effective straining contribution to the magnetic field $H_{fs} \sim L^{-\beta\delta/\nu}$, i.e. $H_{eff} = H + H_{fs}$ such that for $L \to \infty$, $H_{eff} \cong H$.

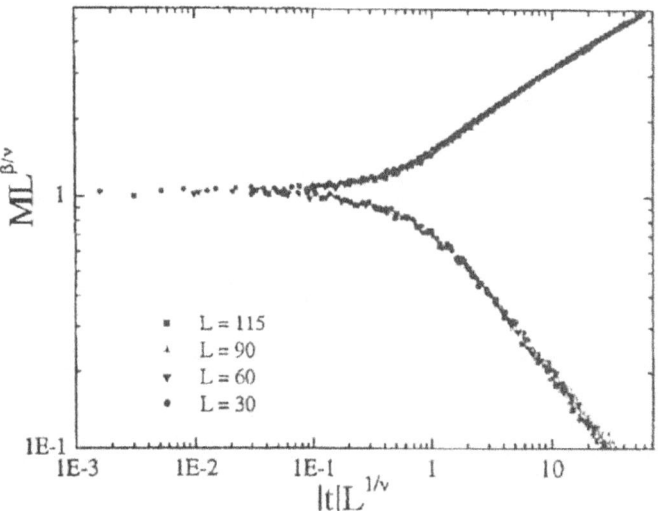

Fig.2 Scaling plot of Monte Carlo data $ML^{\beta/\nu}$ vs $|t|L^{1/\nu}$ for s.c. Ising lattices with linear size $L = 30, 60, 90, 115$. Note that finite size effects for $L = 30$ show up closer to $|t| \to 0$.

Fig. 3 gives scaling plots of the high quality data of 30 isotherms at the vicinity of the Curie temperature pertaining to the insulating ferromagnet CrBr$_3$ measured by Ho and Litster[9,10] which are a classical example of experimental scaling data. We show plots of $\bar{m}$ vs $\bar{h}$ (Fig. 3a) and $\bar{\mu}$ vs $\bar{\tau}$ (Fig. 3b) with $T = T_C = 32.844$ K for two sets of critical exponents, Ising 3d (fractional values) and Ho and Litster (experimental values) summarized in Table 1. Both sets of critical exponents produce good scaling plots in (a) as well as in (b). Apparently the critical exponents of Ho and Litster produce somewhat better scaling plots. The accuracy of the magnetization measurements for the set of isotherms was comparable to that of nuclear magnetic resonance data and it was sufficiently precise to establish the form of the scaling function.

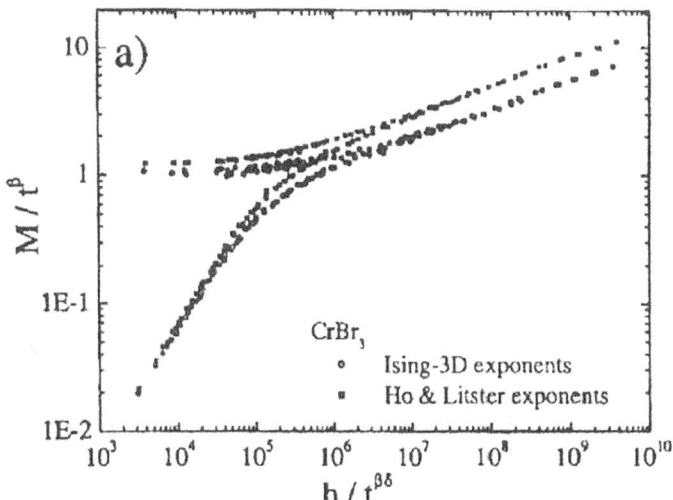

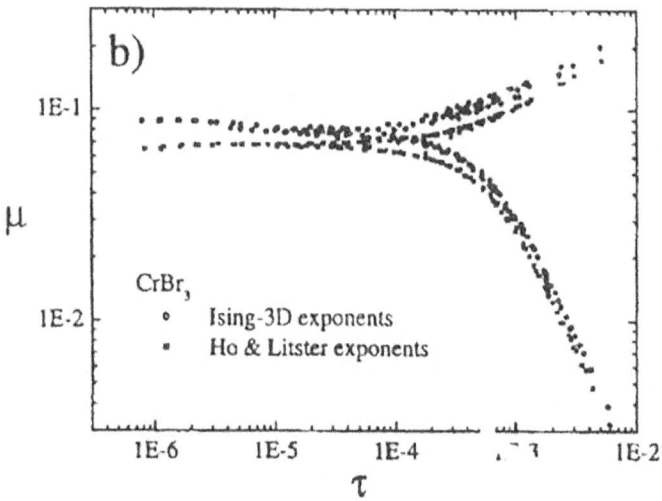

Fig.3 Scaling plots of experimental data for $CrBr_3$. (a) $m \equiv M/|t|^\beta$ vs $\bar{h} \equiv h/|t|^{\beta\delta}$ and (b) $\bar{\mu} \equiv M/h^{1/\delta}$ vs $\bar{\tau} \equiv t/h^{\beta\delta}$. The data are made up of 30 isotherms in the interval $T_C - 0.9$ K $< T < T_C \sim + 6.7$ K. The critical temperature was $T_C = 32.844$ K and the critical exponents used were $\beta = 5/16$ and $\delta = 5$ (Ising 3D) and $\beta = 0.368$, $\delta = 4.28$, as given in Reference [9].

Finally we address the question of the form of the equation of state for 3D Ising lattices in the light of the information provided by the set of isotherms in the vicinity of the critical temperature obtained by the Monte Carlo method in our 70^3 s.c. rewritten as lattice. Fig. 4 gives the plot of $M(h,t)$, Eq. (5)

$$\frac{\bar{h}}{\bar{m}} = f(\bar{m}) = A\left(1 \pm B\bar{m}^z\right)^{(\delta-1)/z} \qquad (10)$$

Table 1 Critical Exponents

	β	δ^{-1}	γ
Ising 3D	5/16=0.3125	1/5=0.2	5/4=1.25
Heisenbeig	0.340	0.208	1.39
CrBr$_3$	0.368	0.233	1.215

Here A can be reduced to unity just by choosing properly the units for the field H. B is a more meaningful coefficient which, in the particular case of a phase transition describable by means of the mean field approximation (such as the phase transition in a uniaxial ferroelectric) is equal to $1/\delta = 1/3$. And z, as pointed out in Reference [8] is given by $z \cong \beta\delta/v = 2.5$ for $T < T_C$, and $z = \beta\delta = 1.562$ for $T > T_C$. Fig. 4 shows the excellent fit obtained by means of Eq. (10) with $(B/A) = 0.102$. The equation of state put in the form given by Eq. (10) is specially good to show directly the most relevant information: (a) the critical isotherm for $T < T_C$ and $T > T_C$, (b) the spontaneous magnetization curve ($T < T_C$) as a vertical line, and (c) the zero field susceptibility ($T > T_C$) as a horizontal line.

The quality of the fit is comparable or better than those obtained with traditional expressions of the scaling function[9,11-17].

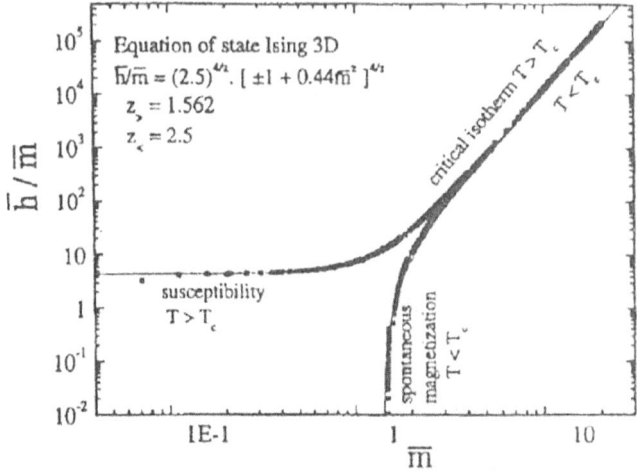

Fig. 4 Optimized fits of the 30 Monte Carlo isotherms to the given Ising 3D equation of state.

Work is in progress to obtain Monte Carlo data in larger 3D Ising lattices at closer field/temperature intervals, and to extend the investigation of scaling plots in the vicinity of the transition to higher dimensionalities Ising 4D, Ising 5D, etc, in order to monitor closely how the approach to mean field behavior takes place. Of course we will be limited to more reduced sizes (smaller L's) as the dimensionality increases, but we have excellent experimental data[18] on a complete set of isotherms in uniaxial ferroelectric TGS to produce excellent scaling plots with T_C = 321.470 and mean field critical exponents $\beta = 1/2$ and $\delta = 3$.

Support from the Spanish MECyD through Grant Number BFM2000-0032 is gratefully acknowledged.

REFERENCES

[1] K. Binder, in Phase Transitions and Critical Phenomena, Vol. 5b (Academic Press, London, 1976).
[2] U. Wolff, Phys. Rev. Lett. 62, 361 (1989).
[3] Metropolis, A. W. Rosenbluth, M. N. Rosenbluth, A. H. Teller, and E. Teller, J. Chem. Phys. 21, 1087 (1953).
[4] J. Garcia and J. A. Gonzalo, Physica A 326, 464 (2003).
[5] J. Garcia and J.A. Gonzalo, [cond-mat/0304056] (to be published).
[6] H. W. J. Blote, L. N. Slichur, and A. L. Talapov, Int. J. Mod. Phys. C 10. 137 (1999).
[7] See f.i. J. M. Yeomans, Statistical Mechanics of Phase Transitions (Oxford University Press, 1992).
[8] M. I. Marques and J. A. Gonzalo, Physica A 267, 165-172 (1999).
[9] J. Ho and J. D. Litster, Phys. Rev. Lett. 22, 603 (1969).
[10] J. Ho and J. D. Litster, Phys. Rev. B 2, 4523-4532 (1970).
[11] A. Arrot and J. E. Noakes, Phys. Rev. Lett. 19, 786 (1967).
[12] M. Vicentini-Missoni, J. M. Levelt Sengers, and M. S. Green, Phys. Rev. Lett. 22, 390 (1969).
[13] S. Milosevic and 11. E. Stanley, Phys. Rev. Lett. 22, 606 (1969).
[14] D- S. Gaunt and C. Domb, J. Phys. C 3, 1442 (1970).
[15] J. A. Gonzalo, Phys. Rev. B 1, 3125 (1970).
[16] S. Milosevic and H. E. Stanley, Phys. Rev. B 6, 986 (1972).
[17] S. Milosevic and H. E. Stanley, Phys. Rev. B 6, 1002 (1972).
[18] J. R. Fernandez del Castillo, B. Noheda, N. Cercccda, J. A. Gonzalo, T. Iglesias, and J. Przcslawski, Phys. Rev. B 57, 805 (1998)

CHAPTER 2.5
EFFECTIVE CRITICAL EXPONENTS AT MINIMAL DIMENSIONALITIES*

Monte Carlo data simulating phase transitions in Ising strips $D \times L$, $(D \ll L)$ with periodic boundary conditions show that $T_C(D) = 0$ for $D \leq D^* \approx 6$ and $0 < T_C(D) < T_C(d=2)$ for $D > D^*$. Regular scaling of $ML^{\beta/\nu}$ vs $|T - T_C| L^{1/\nu}$ is obtained only for $D > D^*$ and the Monte Carlo effective susceptibility critical exponent $\gamma_{\it eff}(D)$ is shown to be well described by $\gamma(d) = \beta(d)[\delta(d) - 1]$ with $d_{\it eff}(D)$ given by $d_{\it eff}(D) \approx 1.5 + (1/200)(D - 6)$ and $\beta(d) = (3d/16 - 1/4)$, $\delta^{-1}(d) = (2d/15 - 1/5)$, which can be understood as valid with $d_{\it eff}(D)$.

Introduction

It is well known[1] that systems at criticality, for instance Ising magnets at $(T \to T_C, H \to 0)$, are usually not only scale invariant but also conformally invariant, i.e., the pertinent local transformations look like combinations of dilations, rotations and translations, but shear distortions are not allowed. This fact has been used to prove that in two dimensions, the Ising critical exponents must be fractional. Monte Carlo methods[2] have been extensively used to characterize phase transitions in Ising systems and are used in this work to study $D \times L$ strips.

*Mª Felisa Martinez, Carlos Garcia & Julio A. Gonzalo, in *Monte Carlo approach to Phase Transitions*, EDS. J.A. Gonzalo, C.L. Wang, Science & Culture, Madrid, 2008)

Recent Monte Carlo[3] data on large three dimensional Ising lattices ($L^3 \approx 1.5 \times 10^6$ spins) appear to indicate that for $2 < d < 4$ (including $d = 3$) the critical exponents are either well defined fractional[4] or indistinguishable from fractional, and that they are determined by compact expressions for $\beta(d)$ and $\delta^{-1}(d)$ (See Fig.1)

In the present work we carry out an empirical investigation of the critical behavior of strips $D \times L$ (with increasing width $D \ll L$) by means of Monte Carlo simulations of the phase transition for long strips with $D = 2, 3, 4...$ up to $D = 52$, using periodic boundary conditions, to ascertain whether the effective critical exponents $\beta(d)$ and $\delta^1(d)$ describing the phase transition are given by the same expressions linear in d which describe well the behavior for higher dimensionalities ($d = 2,3,4$).

Monte Carlo Results

Numerical finite size simulations of the phase transitions in $D \times L$ Ising systems with periodic boundary conditions were performed for $D = 2$ to $D = 52$.

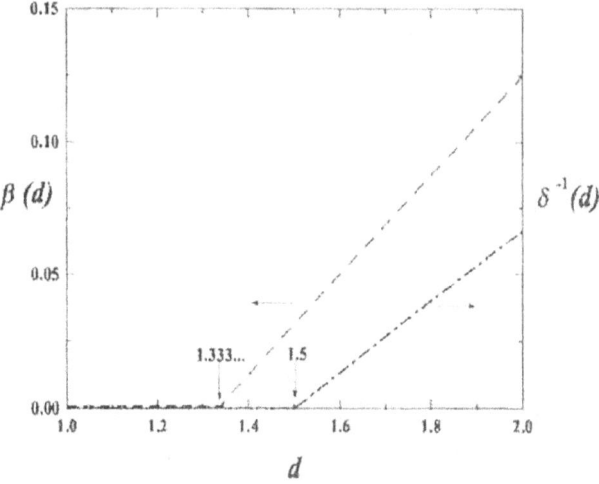

Fig. 1. Critical exponents $\beta(d) = (3d/16 - 1/4)$ and $\delta^{-1}(d) = (2d/15 - 1/5)$ extrapolated from the respective fractional values [$\beta(4) = 1/2$, $\beta(2) = 1/8$] and [$\delta^{-1}(4) = 1/3$, $\delta^{-1}(2) = 1/15$]

Wolff cluster algorithms[5] were used in strips of length $250 \leq L \leq 5000$.

Periodic boundary conditions were used always in the L direction but they had not much effect for large L values as it is to be expected. The thermalization time, relaxation time and number of states were increased steadily until the final results were not appreciably affected by further increases.

140000 Monte Carlo steps per spin for each temperature were taken. To reduce the critical slowing down at $T \approx T_C$, a single cluster Wolff algorithm was used. Initial conditions at any given temperature in a closely spaced set of temperatures ($\Delta T \leq 0.01$) were taken from the equilibrium conditions at the previous temperature.

The critical temperature of strips with width $D \ll L$, which is a non-universal quantity, was determined by means of the Binder cumulant method[2], crossing data for $D \times L$ with data for $D \times 2L$. It was found that for $D \leq 6$ no crossing took place at $T > 0$, and it was checked whether scaling of $ML^{\beta/\nu}$ vs $|\varepsilon|L^{1/\nu}$, with $|\varepsilon| = |T - T_C|$, was possible using $\beta = 0$ and $\nu \leq 1/2$. It was found that no such scaling took place, but, what could be called "one-dimensional" scaling[1], with $|\varepsilon| = e^{-4/T}$ was observed to take place for $D = 2, 3, 4, 5$.

For $D > 6$ the Binder cumulant method did provide noncero transition temperatures given by[2]

$$T_C(D) = T_C(d=2)\left[1 - e^{-m\sqrt{D-D^*}}\right] \qquad (1)$$

with $D^* \approx 6$, and $m \approx 0.353 \pm 0.011$; and regular scaling of $ML^{\beta/\nu}$ vs $|\varepsilon|L^{1/\nu}$ was observed to hold, with $\beta(d)$ and $\delta^1(d)$ evolving smoothly between the respective values for $d = 1.5$ and $d \approx 2$ as specified below.

To describe the evolution of the critical exponents as a function of two-dimensional strip width D we note first that $\beta(d)$, $\delta^{-1}(d)$ and $\gamma(d) = \beta(d)[\delta(d) - 1]$ for $d = 2$, and $d = 4$ are given by fractional values specified by

$$\beta(d) = \left(\frac{3d}{16} - \frac{1}{4}\right) \tag{2}$$

$$\delta^{-1}(d) = \left(\frac{2d}{15} - \frac{1}{5}\right) \tag{3}$$

$$\gamma(d) = \beta(d)\left[\delta(d) - 1\right] =$$

$$= \frac{11}{8} + \frac{45}{192}\left(\frac{2}{2d-3}\right) - \frac{3}{16}\left(\frac{2d-3}{2}\right) \tag{4}$$

For $d = 3$ Monte Carlo data[3] are consistent within narrow statistical error bars with fractional exponents $\beta(d=3) = 5/16$, $\delta^{-1}(d=3) = 1/5$ and $\gamma(d=3) = 5/4$, as given by Eqs. (2), Eq. (3) and Eq. (4) respectively.

Fig. 2 shows plots of χ^{-1} vs $|\varepsilon| = |e^{-4/T}|$ for $D < D^* = 6$ and Fig. 3 shows like plots of χ^{-1} vs $|\varepsilon| = |T - T_C|$ for $D \geq D^*$. It can be seen that $\gamma_{eff}(D)$ diverges at a width $D \approx 6$ from both sides. From these data the evolution of $\gamma_{eff}(D)$ can be determinated with fair accuracy, and it can be used in conjunction with Eq. (4) to determine the effective dimensionality of strips with growing D.

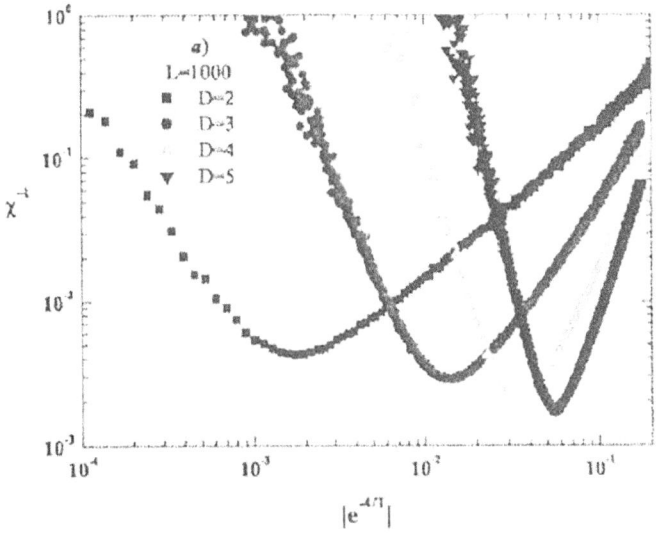

Fig. 2. Inverse susceptibility vs $|\varepsilon| = |e^{-4/T}|$ for $L \times D$ strips with $L = 1000$ and $D = 2, 3, 4, 5$ showing evolution of $\gamma_{\mathit{eff}}(D)$ for $D \leq D^* \approx 6$.

This is done in Fig. 4 where a plot of $d_{\mathit{eff}}(D)$ results in a quiasi-linear dependence of $d_{\mathit{eff}}(D)$ starting at $d_{\mathit{eff}}(D^*) = $ $= 1.5$ (corresponding to $6 - 1 \, (d) = 0$ after Eq. (3) and growing up from this value towards $d_{\mathit{eff}}(L) = 2$, i.e. for rectangular $L \times D$ strips with $D \to L$ (approaching square lattices). $d_{\mathit{eff}}(D)$ therefore, is approximately given by

$$d_{\mathit{eff}}(D) = 1.5 + \left(\frac{1}{200}\right)(D - D^*) \qquad (5)$$

where $D^* \approx 6$.

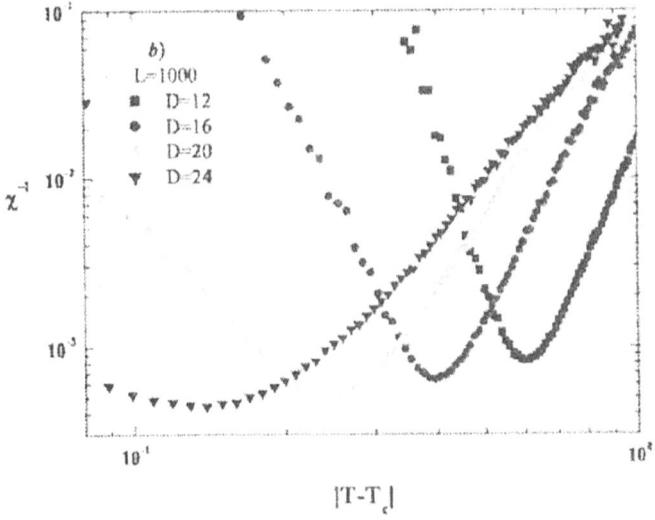

Fig. 3. Inverse susceptibility vs $|\varepsilon| = |T - T_c|$ for $L \times D$ strips with $L = 1000$ and $D = 12, 16, 20, 24$ showing evolution of $\gamma_{eff}(D)$ for $D \geq D^* \approx 6$.

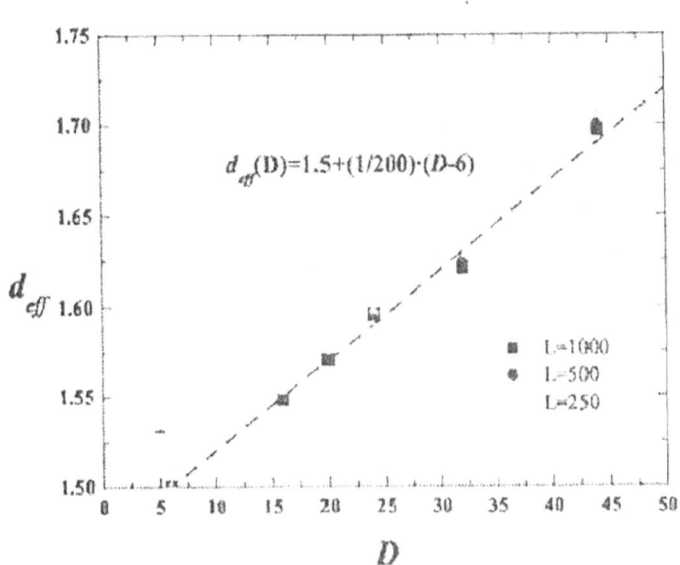

Fig. 4. Effective dimensionality d of $L \times D$ strips as a function of D for

$D \geq D^* \approx 6$ obtained by means of the relationship $\gamma_{\text{eff}}(D) = \gamma(d) = \beta(d)[\delta(d) - 1]$ as given by means of linear extrapolations of $\beta(d)$ and $\delta^{-1}(d)$ from the known fractional values at $d = 2$ and $d = 4$ (see text).

Fig. 5 gives a plot of $d_{\text{eff}}(D)$ which shows directly how this critical exponent blows up as D approaches $D^* \approx 6$, Eq. (4) gives $\gamma(d)$ in terms of $\beta(d)$ and $\delta^{-1}(d)$, as specified by Eqs. (2) and (3). As shown, $\gamma_{\text{eff}}(D)$ is given by

$$\gamma_{\text{eff}}(D) = G + \frac{A}{D - D^*} - B(D - D^*), \quad D > D^* \quad (6)$$

with $G \approx 11/8$, $A \approx 1125/24$ and $B \approx 3/3200$, in good agreement with $\gamma(d)$ given by Eq. (4). For $D < D^*$, i.e. for $D = 2, 3, 4, 5$, $\gamma(d)$ grows up steeply with D and blows up at $D \approx 6$, which corresponds to $\beta(d_c) = 0$ where $d_c = 1.333...$, as given by Eq. (2). $\gamma_{\text{eff}}(d)$ for $D < D^*$, i.e. for the interval $1 < d < d_c$ where $\gamma \approx \beta \cdot \delta \approx (0) \cdot (\infty)$, is undefined, but it can be fitted reasonably well by

$$\gamma_{\text{eff}} \approx C \cdot (D - D^*)^{-1}, \quad D < D^* \quad (7)$$

with $C \approx 4$, which is compatible with $\gamma(d = 1) = 1/2$.

The hyperscaling relationships[1] result in a critical exponent v describing the temperature dependence of the correlation length ξ which is related to the dimensionality d and to other exponents by

$$v^{-1}(d) = \gamma_1(d) = \frac{d}{\beta(d)[\delta(d)-1]} \quad (8)$$

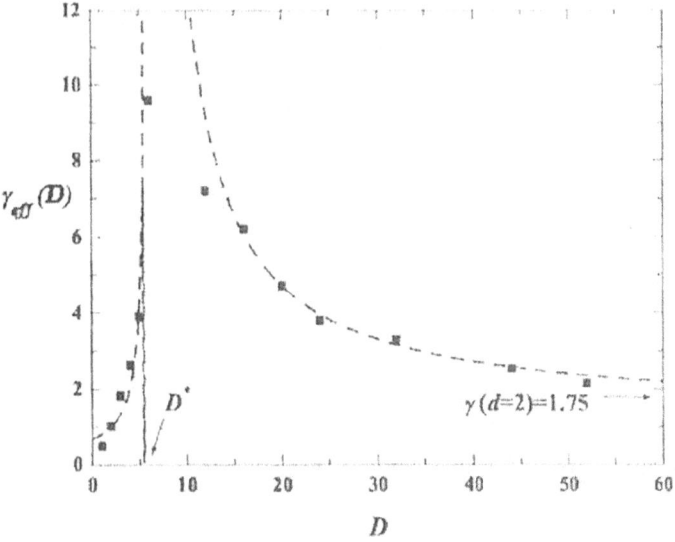

Fig.5. Plot of susceptibility critical exponent $\gamma_{\text{eff}}(D)$ for $L \times D$ strips as a function of D for strips with various widths $D < D^*$ and $D > D^* \approx 6$. The continuous curve for $D > D^*$ is given by $\gamma_{\text{eff}}(D) = G + A(D - D^*)^{-1} - B(D - D^*)$ with $G = 11/8$, $A = 1125/24$ and $B = 3/3200$ in good agreement with $\gamma(d) = \beta(d)[\delta(d)-1]$ as determined by from $\beta(d)$ and $\delta^{-1}(d)$ linearly extrapolated from the respective $d = 2$, $d = 4$ values.

For $D \geq D^*$ the denominator in the right hand side of Eq. (8) can be written as $\{\gamma(d) + 2\beta(d)\} \approx \gamma(d)$ at $d \geq 1.5$. For $D < D^*$ ($1 < d < 1.333..$,) this denominator can be approximated by $\{\gamma(d)\}$, because $\beta(d) = 0$ for $d \leq 1.333...$ Table 1 gives a set of critical exponents, including $(\beta/v)_{\text{eff}}$ and

$(1/v)_{eff}$, to be used below for scaling Monte Carlo data of $ML^{\beta/v}$ vs $|\varepsilon|L^{1/v}$ for strips of various D and $L = 500, 1000$ calculated using periodic boundary conditions.

Fig. 6 and 7 give scaling plots of Monte Carlo data for strips of lenght $L = 500, 1000$ various widths D. Fig. 6 shows results for $D = 3 < D^* \approx 6$, taking $|\varepsilon| = |e^{-4/T}|$, (one- dimensional scaling) and Fig. 7 results for $D = 20 > D^* \approx 6$, taking $|\varepsilon| = |T - T_C|$ (two-dimensional scaling).

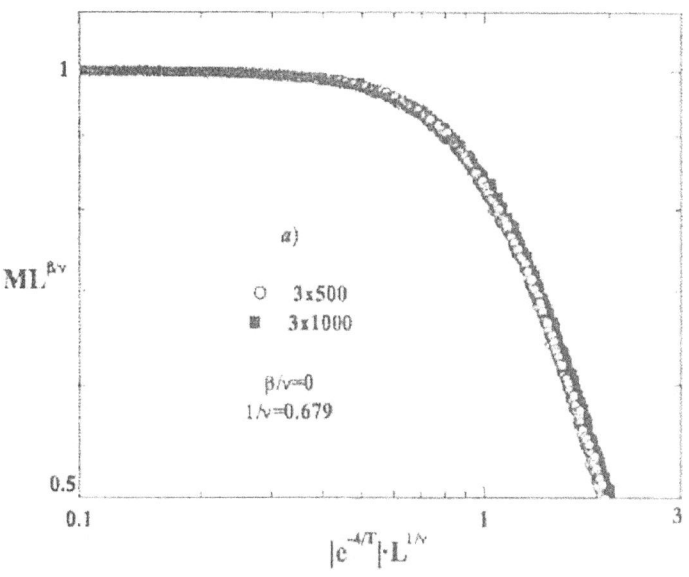

Fig.6. Scaling plots of $ML^{\beta/v}$ vs $|\varepsilon|L^{1/v}$ for strips with $D = 3 < D^*$ taking $|\varepsilon| = |e^{-4/T}|$.

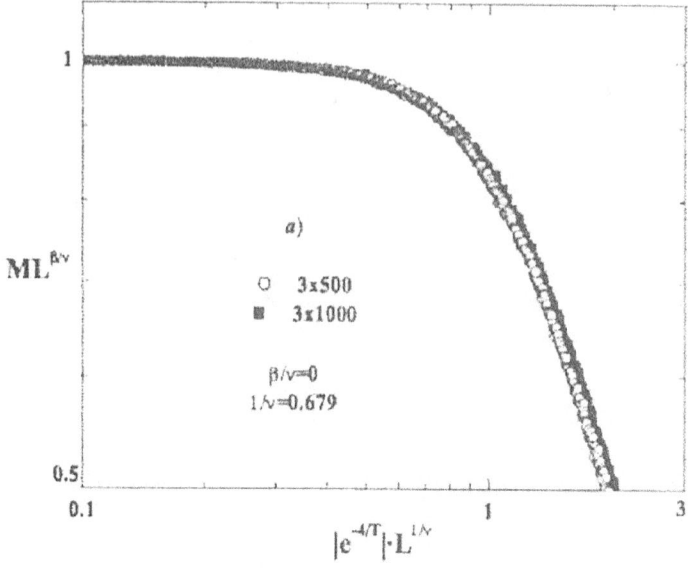

Fig.7. Scaling plots of $ML^{\beta/\nu}$ vs $|\varepsilon|L^{1/\nu}$ for strips with $D = 20 > D^*$ taking $|\varepsilon| = |T - T_C|$.

Concluding Remarks

In summary, our Monte Carlo data properly analyzed show that strips of dimensions $L \times D$, with $D \ll L$, are characterized by a susceptibility critical exponents $\gamma_{\it eff}(D)$ blowing up at $D = D^* \approx 6$, which appears to correspond to a dimensionality $d = 1.333... \leq d \leq 1.5$ which is well determined by the extrapolation of $\beta(d)$ and $\delta^{-1}(d)$ given by the same linear expressions which describe well the dimensionality dependence for $d = 2, 3, 4$. This lends empirical support to the theoretical contention that the renormalization transformations, which are not only scale invariant but also conformally invariant transformations, imply fractional exponents for Ising lattices of geometry intermediate between one dimensional and two dimensional.

Table 1. Ising critical exponents for dimensionalities $1 \leq d \leq 2$

d_{eff}	d_{eff}	d_{eff}	d_{eff}	d_{eff}	d_{eff}
1.0	0	0	1/2	2	0
1.1	0	0	(0.71)	(1.54)	0
1.2	0	0	(1.25)	(0.96)	0
1.3	0	0	(5.00)	(0.26)	0
1.333...	0	0	00	0	0
1.5	0.031	0	00	0	0
1.6	0.050	0.013	3.700	0.421	0.021
1.7	0.069	0.027	2.516	0.642	0.044
1.8	0.087	0.040	2.100	0.791	0.069
1.9	0.106	0.053	1.887	0.905	0.096

It may be pointed out that exact information for the largest eigenvalues of the transfer matrix of the system in small width strips (perhaps up to $D < 20$ or so with infinite length) can be obtained using f.i., the Lanczos algorithm, but, in principle, this does not result in useful information about the effective exponents, which are the ones which determine the observable behaviour of $M(T)$ close to, but not arbitrarily close to the transition.

Acknowledgments: Support from the DGICyT for grant BFM2000-0032 is gratefully acknowledged.

REFERENCES

[1] Yeomans, *Statistical Mechanics of Phase Transitions* (Oxford University Press, 1992).
[2] See f.i. K. Binder, Phys. A 98 *Phase transitions and Critical Phenomena*, Vol.5B (Academic Press: London 1976).
[3] J Garcia and Julio A. Gonzalo, *Physica A* (2003, inpress).
[4] See f.i. J Kaupuzs, *Annalen*, Phys. 10, 299-331 (2001).
[5] U. Wolff, *Phys. Rev. Lett.* 62, 361 (1989).

Felix Bloch (1905 – 1998)

CHAPTER 2.6
FERROELECTRIC DIPOLE WAVES.

Low temperature dipole-wave-like excitations in ferroelectrics which are analogous in certain respects to ferromagnetic dipolar spin waves (see, for example, Herringa and Marrenga 1975), might be expected on the grounds that, slightly above zero temperature, collective-wave-like deviations of the elementary dipoles from maximum alignment would require much less energy than the reversal of single dipoles would. Ordinary acoustic phonons can coexist in a non-interacting way with low-energy spin-wave-like excitations in solids possessing a spontaneous polarization. Consider a complex crystal in which lattice points are occupied by rigidly bound groups of atoms instead of single ions. Suppose that these groups of atoms possess permanent electric dipole moments and that, below a certain transition temperature, there is a non-zero spontaneous polarization and, therefore, an associated spontaneous local field. To find rigorously the ground state and the full spectrum of excited states of this system would demand the solution of the many-body problem in one of its most difficult forms. One may reasonably argue, however, that under favorable conditions the dipole wave excitations can be considered as low-energy internal vibrational modes and then can be treated separately from the normal low-temperature acoustic phonons, which involve vibrations of the tightly bound group of atoms making up a unit dipole. Intuitively, the energy required to produce long-wavelength periodic deviations of the rigid dipoles from perfect alignment in the direction opposite to the field goes to zero as q goes to zero. We shall see below that explicit separate consideration of the long-range dipole-dipole forces acting on a given elementary dipole gives rise to $\omega_q \simeq S_{sw}(q)q^2$ which in turn leads to $T^{3/2}$ contributions to the low-temperature specific heat and a low temperature change in the spontaneous polarization. Ordinary acoustic phonons give rise independently to a Debye T^3 contribution to the specific heat and to a zero net contribution to the change in spontaneous polarization, (It is not uncommon to find reports in the literature of crystalline systems, e.g. some ammonium compounds, in which certain degrees of freedom are "frozen" at $T = 0$ and then begin to be evident when the temperature is raised, as librations of groups of atoms which were formerly behaving as single point-like masses). This effect, which shows up as a temperature-dependent effective Debye tenperature, is broadly ascribed to anharmonic behaviour. However, it is well known that high-energy internal modes of molecular units in a solid are often relatively unaffected by the presence of acoustic modes of very different energies. In the same way, it is perfectly conceivable that the dipole waves, which are considered as very-low-energy internal librational modes, behave as non-interacting modes with respect to acoustic phonons of energy $\omega_q \simeq S_{ph}(q)q$, at least for temperatures low enough for the relevant (excited) wave vectors q to be such that $q < S_{sw}/S_{ph}$. This is equivalent to saying that short-range forces may remain near equilibrium while long-range dipole- dipole forces (responsible for the dipole waves) are playing an active role. Of course, a definitive check on the actual existence of excitations which behave as $\omega_q \simeq S_{sw}(q)q^2$ must wait until systematic neutron scattering data well inside the Brillouin zone are available (at low and high temperatures). In the ground state ($T = 0$), dipole alignment should be maximum, but not perfect, because of zero-point motion. The situation is partly analogous to that encountered in dipolar (not exchange) ferromagnets. The main difference between ferroelectric and (dipolar) ferromagnetic systems lies in the well-defined character of the atomic magnetic dipole moment in the latter, in contrast with the ill-defined electric dipole moment which makes up the complex primitive unit cell in most ferroelectrics. A previous attempt to attack this problem (Gonzalo 1978) was clearly insufficient and incomplete, partly because not enough experimental evidence was available. A more systematic calculation is presented here, and its results are compared at the end with available experimental data on TGS. A systematic calculation of basic properties in ferroelectric at low temperatures is interesting in its own right, even though comparison with experiments is difficult because of the somewhat conflicting results obtained by different researchers (see discussion below).

To investigate the anomalous low-temperature behavior in uniaxial ferroelectrics, we shall consider only the long-range dipole interactions, since short-range interactions will give rise to normal behavior (Debye behavior of the specific heat) which can be treated separately from the former.

A simple dipole-dipole interaction Hamiltonian $\left(H \equiv H_{dipole-dipole}\right)$ of a system of N elementary electric dipole moments in a uniaxial ferroelectric crystal under zero external field can be written as

$$H = -\sum_{ll'} \frac{1-3\cos^2\theta_{ll'}}{|l-l'|} \mu_l \mu_{l'} = \sum_{ll'} J_{ll'} \mu_l \mu_{l'} \qquad (1)$$

where $\theta_{ll'}$ is the polar angle between $l-l'$ and the polar axis z, parallel to the spontaneous polarization. The dipoles are assumed to be located at regularly spaced fixed points in the crystal lattice and almost perfect alignment is expected in the ferroelectric phase at low temperatures. If we leave aside the short-range forces, the net restoring force due to all neighboring dipoles on a given dipole can be represented adequately by the spontaneous local field E_s acting on the dipole. The usual mean-field approximation consists in assuming that every dipole in the crystal is under the influence of an effective field $E_{eff} = E + E_s = E + \beta P$, where E is the external field, E_s the cooperative "spontaneous" field and β a constant. The last assumption is justified only at a temperature sufficiently lower than the transition temperature T_C so that lattice distortions are small, the unit cell volume remains almost constant, and the value of the order parameter is very close to unity. The behavior of the system may change substantially when appreciable anisotropic thermal expansion and new internal degrees of freedom begin to enter the picture.

At any temperature higher than 0 K, every $\mu_l(t)$ undergoes, under the influence of the spontaneous local field E_s, a pseudo-regular precession (see, for example, Sommerfeld 1964), which can be thought of as a regular precession around E_s accompanied by a small nutation of higher frequency, which might perhaps be associated with an "uncertainty" in the μ_z component, parallel to E_s.

We may anticipate that this motion is compatible with a low-temperature specific heat which is proportional to $T^{3/2}$, because we have two kinetic degrees of freedom (motion of the tip of μ_l towards $z \| E_s$, and motion perpendicular to z) plus one potential energy degree of freedom, defined by the angle α between μ_l and E_s. The angular momentum $L_l(t)$ associated with the motion of $\mu_l(t)$ is perpendicular to $\mu_l = qd$ and to the velocity of its tip, and it is undergoing a regular precession.

Therefore, from

$$\mu_l \mu_{l'} = \mu^2 \cos \alpha_{ll'}, \qquad (2)$$

$$L_l L_{l'} = L^2 \cos \alpha_{ll'}, \qquad (3)$$

we obtain, dividing Eqs. (2) and (3),

$$\gamma^2 = \mu^2/L^2 = \mu_l \mu_{l'}/L_l L_{l'} = \text{constant} \tag{4}$$

where, through the virial theorem,

$$\varepsilon = 2T_k = 2V, \quad T_k = V, \quad L^2/2I = \mu E_s \cos\theta \tag{5}$$

Here T_k is the kinetic energy, V is the potential energy, $I = Mr^2$ is the moment of inertia of the dipole, M is its total mass and θ is the angle between $\mu = qd$ and E_s. For θ small, and taking into account that within the mean-field approximation (see, for example, Gonzalo 1978, and references therein)

$$\mu E_s = \beta N \mu^2 = k_B T_C \tag{6}$$

we obtain, after substituting Eq.(6) into Eq. (5) and Eq. (5) into Eq. (4),

$$\gamma^2 \simeq (d/r)^2 \left(q^2/2Mk_B T\right) \tag{7}$$

Consequently, we can substitute

$$\mu_l \mu_{l'} = \gamma^2 L_l L_{l'}, \tag{8}$$

in Eq. (1), leaving it in terms of angular momenta instead of elementary dipole moments. We can now use directly the formalism of angular momentum quantisation (see, for example, Taylor 1970) for a system of N quantised processing dipoles to obtain

$$\omega_q = \gamma^2 h \sum_{l'} J_{ll'} \{1 - \exp[-iq(l-l')]\} = \\ = q\gamma^2 h \sum_{l'} J_{ll'} \sin^2\left[\frac{1}{2}q(l-l')\right] \tag{9}$$

We can use $\langle n_q \rangle = [\exp(\hbar\omega_k/k_BT)-1]^{-1}$ (thermal average) in H to obtain

$$H = H_0 + \sum_q \hbar\omega_q n_q =$$
$$= H_0 + \sum_q \hbar\omega_q \left[\exp(\hbar\omega_q/k_BT)-1\right]^{-1} \quad (10)$$

(Note that a uniaxial ferroelectric is equivalent, for present purposes, to a spin $\frac{1}{2}$ ferromagnet and that in the intervening transformations we have neglected interactions between dipole waves.)

To calculate the low-temperature contributions of the dipolar waves to specific heat and spontaneous polarization it is convenient to rewrite H as follows:

$$H - H_0 = \sum_q \hbar\omega_q \langle n_q \rangle =$$
$$= \sum_q \hbar \left[\gamma^2 \hbar\alpha(q) J_0 \sin^2\left(\frac{qd_c}{2}\right)\right]\langle n_q \rangle = \quad (11)$$
$$= \sum_q \alpha(q) k_B \theta \sin^2\left(\frac{qd_c}{2}\right)\langle n_q \rangle$$

where

$$\alpha(q) = \left\{\sum_{l'} J_{ll'} \sin^2\left[\frac{q(l-l')}{2}\right]\right\} \bigg/ \sum_{l'} J_{ll'} \sin^2\left(\frac{qd_c}{2}\right)$$

$$J_0 = \sum_{l'} J_{ll'}, \quad d_c = v_c^{1/3} = N^{-1/3},$$

$$\theta = h^2\gamma^2 \sum_{l'} J_{ll'}/k_B \qquad (12)$$

These are simple definitions which, for the time being, involve no approximations. The $\alpha(q)$ are dimensionless parameters, in principle of order unity, d_c is a characteristic length related to the unit-cell volume ($v_c = N^{-1}$) and θ is a characteristic temperature, related to the strength of the dipolar interaction in the geometry of the crystal lattice.

The low-temperature specific heat contribution due to dlpole-wave-like excitations is given by

$$\Delta C_v = \frac{\partial}{\partial T}\left(\sum_q n_q h\omega_q\right) =$$

$$= \frac{1}{8\pi^3}\frac{\partial}{\partial T}\cdot\left(k_B T \int \frac{\alpha(q)(\theta/T)\sin^2(qd_c/2)}{\exp\left[\alpha(q)(\theta/T)\sin^2(qd_c/2)\right]-1}d^3q\right) \qquad (13)$$

This can be written, using $x = qd_c/2$, as

$$\Delta C_v =$$

$$= \frac{\partial}{\partial T}\left(\frac{2^3}{8\pi^3}Nk_B T\int_{BZ}\frac{\alpha_q(\theta/T)\sin^2 x}{\exp\left[\alpha_q(\theta/T)\sin^2 x\right]-1}\Omega x^2 dx\right) \qquad (14)$$

where, for a given range of x, α_q is an effective average of $\alpha(q)$ and $\Omega \leq 4\pi$ is an effective integration solid angle. Since we are interested mainly in bulk properties, the average takes care of the characteristic directional dependence of ω_q. We may assume a cylindrical Brillouin zone (BZ) for a uniaxial ferroelectric ("pill box" shaped for TGS) and then perform an approximate integration in two steps.

(i) For $0 \leq q \leq q_{D1}$, the central sphere with radius equal to the half-height of the box, $\Omega = 4\pi$, $\sin^2 x \simeq x^2$, and the first part of the integral takes the form

$$\int_0^{x_{D1}} \alpha_q(\theta/T) x^4 \left\{\exp\left[\alpha_q(\theta/T)x^2\right] - 1\right\}^{-1} dx$$

(ii) For $q_{D1} \leq q \leq q_{D2}$, the part of the cylindrical BZ outside the central sphere, where q_{D2} is the half-diameter of the box, $\Omega = 4\pi(q_{D1}/q)$, $\sin^2 x \simeq 1$ (if the ratio q_{D2}/q_{D1} is not abnormally large), and the second part of the integral takes the form

$$x_{D1} \int \alpha(\theta/T) x \left\{\exp\left[\alpha_q(\theta/T)\right] - 1\right\}^{-1} dx$$

Using these rather rough approximations (which are needed to make comparisons below with experimental data) and substituting the appropriate results for the definite integrals, one obtains

$$\Delta C_v = B_1 T^{3/2} + B_2 \left(\theta_D^C/T\right)^2 \exp\left(-\theta_0^C/T\right) \tag{15}$$

where

$$B_1 = 0.904 \, Nk_B \left(2\alpha_0^C \Theta\right)^{-3/2} \tag{16}$$

$$B_2 = \left(2^3/2\pi^2\right) Nk_B r_C \Delta \tag{17}$$

$$\Theta_D^C = 2\alpha_D^C \Theta = 2\alpha_D^C h\gamma^2/k_B \tag{18}$$

Here α_0^C is the mean value of α_q for $0 \leq q \leq q_{D1} = 2x_{D1}/d_c$ and α_D^C the mean value of α_q for $q_{D1} \leq q \leq q_{D2}$. The superscript C refers to specific heat, to differentiate the α^C from the corresponding α^P for spontaneous polarization to be introduced below. Note that the q dependence of the α_q for simple lattices is rather smooth, especially for $q \to 0$ (see, for example, Herringa and Marrenga 1975, and references therein). The parameters $r_C \equiv \{\exp(\alpha_D^C \Theta/T)/[\exp(\alpha_D^C \Theta/T)-1]\}^2$, very close to unity as long as $\Theta_D^C \equiv \alpha_D^C \Theta/T$ (which is the case for the experimental data to be considered later), and $\Delta \equiv (\pi/2)[1-(q_{D1}/q_{D2})^2]$, which is temperature independent, are introduced to simplify the notation.

The low temperature change in spontaneous polarization from its maximum value at $T = 0$ K Is given by

$$\Delta P^2 = \sum_l \Delta\mu_l^z = \sum_l \gamma \Delta L^z = \gamma h \sum_l n_l = \gamma h \sum_q n_q =$$

$$= \frac{\gamma h}{8\pi^3} \int_{BZ} \frac{1}{\exp(h\omega_q/k_B T)-1} d^3q \tag{19}$$

The same procedure used to evaluate ΔC leads now to

$$\Delta P^z = A_1 T^{3/2} + A_2 \exp(-\Theta_D^C/T) \tag{20}$$

where

$$A_1 = 0.469 \, N\gamma h \left(2\alpha_0^P \Theta\right)^{-3/2} \tag{21}$$

$$A_2 = \left(2^3/2\pi^2\right) N\gamma h r_P \Delta \tag{22}$$

$$\Theta_D^P = 2\alpha_D^P \Theta = 2\alpha_D^P h\gamma^2/k_B \tag{23}$$

Here, again, α_0^P and α_D^P are mean values for the integration intervals $0 \leq q \leq q_{D1}$ and $0 \leq q \leq q_{D2}$, respectively. (Note that the approximate integrand in the first interval, which contains x^2 in the numerator instead of x^4, is slightly different from the previous integrand, while the approximate integral in the second interval is exactly the same as before). The parameters $r_P \simeq r_C$ and Δ have the same meanings as those introduced in calculating ΔC_V.

Combining Eqs.(16)-(18) and Eqs. (21)-(23). we obtain

$$A_1/B_1 = 1.92\left(\alpha_0^P/\alpha_0^C\right)^{-3/2} \gamma h/k_B \tag{24}$$

$$A_2/B_2 = \left(r_P/r_C\right)\gamma h/k_B = \gamma h/k_B \tag{25}$$

$$\Theta_D^P/\Theta_D^C = \alpha_D^P/\alpha_D^C = 1; \quad k_B\Theta_D = 2\alpha_D h\gamma^2 \tag{26}$$

These simple relationships connect low-temperature spontaneous polarization and specific heat behaviour through the ratio $\gamma h/k_B \simeq \delta \mu/k_B$.

As mentioned in 1, the basic objective of this work is to present a systematic theoretical description of the physical consequences of spin-wave-like excitations in ferroelectric crystals, leaving aside a definitive statement on whether they can be identified with the observed low-temperature anomalies in some ferroelectrics (Lawless 1976a,b). Alternative explanations, such as impurity effects, have been suggested (Kirkpatrick and Varma 1978) for the anomalous $T^{3/2}$ specific heat behavior reported for some ferroelectrics. This latter work, however, does not discuss either the exponential dependence with $1/T$ reported at a slightly higher temperature range, or spontaneous polarization data apparently following the same pattern (Vieira *et al*. 1978) which was published shortly afterwards. Further experimental work (Lawless 1981) suggested a surface-to-volume ratio of the specific heat $T^{3/2}$ term coefficient not confirmed by latter work (Foot and Anderson 1985), in which an excess observed specific heat contribution behaving as T^n, with $1 < n < 2$ was also reported. In contrast, very-low-temperature specific heat measurements in various ferroelectrics (Grimm *et al*. 1984, Villar *et al*. 1986) using ultrapure samples seem to support no excess low-temperature contributions, or extremely small ones, above the normal T^3 Debye background. In connection with the last data, it may be pointed out that the only exception among the five crystals studied seems to be TGS, for which data are shown to be reduced by a factor of 10. If one plots the points corresponding to $T < 1$ K on the same scale as the data for the other samples and then extrapolates to $= 0$ K, the result is a $T^{3/2}$ contribution slightly lower (by less than 30%) but very similar to that originally reported (Lawless 1976a,b). In summary, while the experimental situation is not yet clear, one may say that at least for TGS after all the available evidence (from both low-temperature specific heat and spontaneous polarization data) has been considered, it is not unreasonable to assume the presence of a $T^{3/2}$ contribution followed by an exponential contribution. In the following, we have used data of Lawless (1976a, b) for B_1 in TGS (assuming that it is at least within an order magnitude of the true experimental value) as well as the data of Vieira *et al.* (1978) for A_1

Low-temperature specific heat data from ferroelectric TGS single crystals (Lawless 1976a, b) (in cgs units per unit volume) give

$$B_1 = 12.9, \quad B_2 = 9.1 \times 10^5, \quad \Theta_D^C = 67 \text{ K} \quad (27)$$

and corresponding spontaneous polarization data (Vieira *et al.* 1978) (in cgs esu units) give

$$A_1 = 0.99 \times 10^{-4}, \quad A_2 = 28, \quad \Theta_D^P = 66 \text{ K} \quad (28)$$

Then, from Eqs. (24), (25) and (26), respectively, we obtain three independent estimates of γ, in cgs esu units:

$$\gamma = 5.2 \times 10^5 \left(\alpha_0^P / \alpha_0^C\right)^{3/2}, \quad \gamma = 4.0 \times 10^6,$$
$$\gamma = \left(1/\alpha_D\right)^{1/2} \cdot 2 \times 10^6 \quad (29)$$

The internal consistency of these data is excellent, since the α_0 and α_D are dimensionless numbers of the order of unity.

Now, let us return to the theoretical expression for γ in Eq. (17). Substituting in some values, we can check whether this theoretical value for γ based on the motion of molecular dipole units is or is not in agreement with the experimental values in Eqs. (29). Let us use, again in cgs esu units,

$$q \simeq e = 4.8 \times 10^{-10},$$

$$N = v_C^{-1} = 0.155 \times 10^{22},$$

$$d = \mu/e = P_{so}/Ne \simeq 1.7 \times 10^{-8}$$

$$r \simeq v_C^{1/3} \simeq 8.6 \times 10^{-8},$$

$$M = Mm_P = 321 \cdot 1.67 \times 10^{-24} \text{ and}$$

$$k_B T_C = 1.38 \times 10^{-16} \cdot 322.$$

Substituting these values in Eq. (7), we obtain

$$\gamma_{calc} \simeq 13 \times 10^6 \qquad (30)$$

which is in fairly good agreement with γ_{exp} (Eqs.(29)).

It has bee shown that spin-wave-like elementary excitations in ferroelectrics lead to extra low-temperature contributions to the specific heat of the form

$$\Delta C_v(T) = B_1 T^{3/2} + B_2 \left(\Theta_D^C/T\right)^2 \exp\left(-\Theta_D^C/T\right)$$

and low-temperature changes in the spontaneous polarization of the form

$$\Delta P_s(T) = A_1 T^{3/2} + A_2 \exp\left(-\Theta_D^P/T\right)$$

where the coefficients in both expressions are related to each other by

$$A_1/B_1 = 1.92 \left(\alpha_0^P/\alpha_0^C\right)^{-3/2} \gamma h/k_B$$

$$A_2/B_2 = \gamma h/k_B$$

$$\Theta_D^P/\Theta_D^C = 1; \quad k_B\Theta_D = 2\alpha_D h\gamma^2$$

The low-temperature data available for TGS are in fair agreement with these expressions.

REFERENCES

[1] M.C. Foot and A.C. Anderson, *Ferroelectrics* 62, 11 (1985)
[2] J.A. Gonzalo, *Ferroelectrics* 20, 159 (1978)
[3] H. Grimm, R. Villar and E. Gmelin, *Proc. 17th Int. Conf. Low Temperature Physics* ed. U. Eckern *et al.* (Amsterdam: North-Holland) p. 1377
[4] J.R. Herringa and M. Marrenga, *Physica A* 38, 371 (1975)
[5] S. Kirkpatrick and C.M. Varma, *Solid State Commun.* 25, 821 (1978)
[6] W.N. Lawless, *Phys. Rev. Lett.* 36, 478 (1976a); *Phys. Rev. B* 14, 134 (1976b); *Phys. Rev. B* 23, 2421 (1981)
[7] A. Sommerfeld, *Mechanics* (New York: Academic) (1964)
[8] P.L. Taylor, *A Quantum Approach to the Solid State* (Englewood Cliffs, NJ: Prentice-Hall) (1970)
[9] S. Vieira, C. de las Heras and J.A. Gonzalo, *Phys. Rev. Lett.* 41, 1822 (1978)
[10] R. Villar, E. Gmelin and H. Grimm, private communication (1986)

3. Selected Global Problems

3.1 Introductory Considerations.

3.2 On a Rate Equations Approach to World Population Trends.

3.3 On World Population Slowing Down.

3.4 On the future World Population Decline.

3.5 At which side will be history at the end of the 21st century?

CHAPTER 3.1
INTRODUCTORY CONSIDERATIONS

In Section 3, the last Section of this book, preceded by this Introductory Chapter 3.1, elementary considerations on a few global contemporary problems are given. In 3.2 the future of world population, which more than half a century ago was evaluated by H. von Foester et al. in *Nature* as actually going to infinity in the year 2026 ("Doomday November 13, 2026) is shown to be described well by a **rate equation** similar to that used in condensed matter physics to describe the time evolution of two level systems under specific changes, birth rate, death rate or both. It is shown that the very large increase in world population form 1950 to 2010 is slowing down clearly. World population will begin to go down about 2045 contrary to the catastrophic Malthusian expectations anticipated in the early sixties. These expectations were not well founded. The increase in world population was not due to any increase in average fertility (inexistent since 1950) but to a substantial increase of life expectancy, first in the more developed countries and then in the whole world. The world is not overpopulated. Only very large cities are overpopulated.

In 3.3 the slowing down of population in Europe and America since the middle sixties of last century is examined in historical perspective. It is pointed out that first in Europe and then, at both sides of the Atlantic, the juridical order previously based in classical Greco-Roman wisdom, the Decalogue and the Gospels underwent a cultural, social and moral decomposition. This decomposition took place, unexpectedly, in striking coincidence with the fall of Soviet Communism. Christopher Dawson one of the greatest historians of last century saw it coming and did a very instructive analysis of the ruthless materialism which degrades de human spirit in our times. According to him, however the wave that is eroding the traditional moral and religious principles of the Western Christian Civilization will eventually slow down. Christian tradition has only suffered a great loss in social influence and in intellectual prestige. To hold up clear theological principles has become uphill, countercurrent in our times. But sooner or later, the materialistic wave will run out, and man will recover the sense of the spiritual and the thirst for tracendent realities which distinguishes him from the primates and from the lower animals guided mainly by their instincts.

Finally, in 3.4 it is examined how future historians are likely to judge the end of the 20th century. As we know, civilizations, make up of cultural components, large or small, very similar or substantially dissimilar can be born, grow and break up. After discussing ongoing trends in population growth or lack of it, rent per capita and religious convictions in the Europe, America, Asia and the Islamic world, and after examining separatism and centrifugal tendencies in some European countries short term proyections not very optimistic can be made about what to be expected for the middle of the present century. In any case, the competent considerations of Christopher Dawson must be kept in mind.

CHAPTER 3.2
ON A RATE EQUATIONS APPROACH TO WORLD POPULATION TRENDS *

Introduction

The UN data on world population, fertility rates, birth and death rates provide abundant and reliable information to investigate present population trends, and to make guesses about the future. These data can be complemented with vital statistical data from the UN Department of Economic and Social Affairs.[1]

In what follows we introduce population rate equations (Dekker, 1962; Loudon, 2000; Gonzalo et al., 2002) appropriate to describe the evolution of the population of a two level system in terms of the known birth rate $(BR = r_b)$ and death rate $(DR = r_d)$, under the assumption that total biomass is approximately conserved. Global biomass would play the role of a universal constant that somehow provides an upper limit to population, at least in the short-medium term. Obviously global biomass can change within certain limits depending on environmental conditions but it is not completely elastic. In this paper we use a rate equations approach (Gonzalo, 2010) to describe world population dynamics which involve a conservative principle. Within a human generation (about 25 years) the world total biomass may be expected to change at most a factor of 2 to 3 –except if a nuclear war,

---

* Julio A. Gonzalo, Felix F. Muñoz and David Santos, *Simulation* (Transactions of The Society for Modeling and Simulation International), February 2013, Volume 89 Number 2.

global pestilence or the collision of a comet with our planet take place.

The Earth has a biosphere capable of sustaining life: plants, animals and men. It has been estimated that the total amount of living matter on Earth is about 1.4×10^{15} Kg according to the World Atlas of Biodiversity (Groombridge and Jenkins, 2002).[2] Men, like other higher animals, live on a diet of carbohydrates, fats and proteins. As it is well known, in the last instance, the chain of life in our planet begins with photosynthesis by the green plants. According to FAO (see Vian Ortuño 1994) the distribution of vegetal biomass is roughly the following: 46% forests and lower wild vegetation; 10% pastures and steppes; 1.8% deserts; 6.2% crops; 2.9% biomass in continental waters; 33% biomass in the oceans. At present the ratio of human to total biomass is roughly

$$\frac{\text{Human Biomass}}{\text{Total Biomass}} \approx \frac{60 \times 6.8 \times 10^9}{1.4 \times 10^{15}} \approx 2.9 \times 10^{-4}$$

The consumption (and large-scale waste) of energy is of course an important factor affecting world population. However, it should be kept in mind that converting directly 1/6000 of the radiant energy coming from the Sun would provide sufficient electrical or chemical energy to satisfy present world energy needs. Fusion energy research may someday contribute effectively to provide abundant, safe and clean energy.

Under the above mentioned assumption, the solutions of rate equations are shown to simulate quantitatively well the actual time evolution of the world population $P(t)$ in the second part of the twentieth century, and to provide reasonable grounds for estimating quantitatively the short term evolution of world population. The main result of this paper is that a linear extrapolation of the actual data up to 2020 suggests a population decrease in the world at about mid-century. This result is consistent –at least partially- with UN, World Bank and IIASA projections for the rest of the century (assuming the medium scenario).[3]

As far as we know rate equations have not been previously used to determine world population trends. A main advantage of this approach is that it needs very few numerical parameters to simulate population trends in comparison to other alternative approaches.[4]

The paper is organized as follows: in section 2 we derive the rate equations that we use to describe and predict world population in the 21st Century; section 3 offers a brief analysis of World population data from 1950 to 2010; section 4 presents the results of applying rate equations to describe quantitatively world population trends; and finally, section 5 is the conclusion.

Rate Equations

We consider a certain step change (up or down) in population associated with a correlative change (up or down) in human biomass $\Delta M(Kg)$ such that the final change in population (after the step is over) is given by

$$\Delta N \cong \Delta M \ (Kg)/60 \ (Kg) = N_2(t) + N_1(t) \qquad (1a)[5]$$

Where $N_2(t)$ is the number of individuals (men and women) actually alive at time t, and $N_1(t)$ the number of individuals potentially alive at the same time. $N_1(t)$ corresponds, therefore, to the amount of biomass potentially convertible into human mass at that time. Time t is in the interval (beginning) $t_i \leq t \leq t_f$ (end), where t_f is sufficiently larger than the relevant characteristic time τ^* for the step (up or down). We can assume ΔN to be in the range $10^9 < N < 10^{11}$, in principle, and N_{RL} (the background replacement level population at $t < t_i$) to be of the same order of magnitude.

During the transition between $N(0) = N_{RL}$ (original replacement level) to $N(t_f) = N_{RL} + \Delta N(t_f)$ (final replacement level) the excess human biomass is distributed between the two levels in such a way that always

$$N_2(t) + N_1(t) = \Delta N(t_f) \tag{1b}$$

At any given time there is a certain birth rate, $BR = r_b$, governing the transition of human biomass from level 1 to level 2, and a certain death rate, likewise governing the transition from level 2 to level 1.

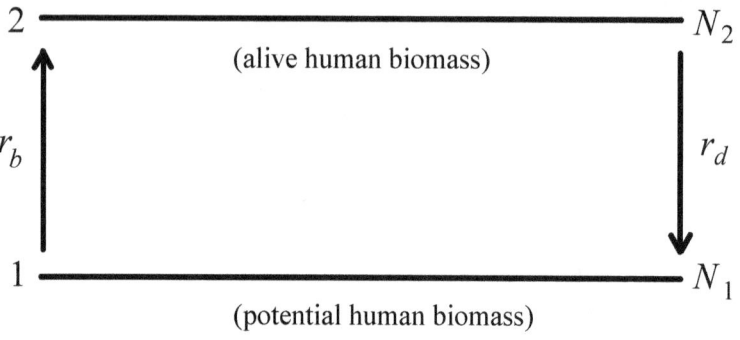

Figure 1: The world as two level system to describe step (+/-) over a certain population background (Replacement Level).

We can define the respective transition probabilities (say

per 100 or 1000 persons) per unit time as

$$p_{12} = r_b = r_0 e^{\alpha} \quad \text{(birth rate)} \tag{2}$$

$$p_{21} = r_d = r_0 e^{-\alpha} \quad \text{(death rate)} \tag{3}$$

where $r_0 \equiv p_0 e^{-\alpha_0}$ could be viewed as the product of an attempt frequency p_0 (inverse of a natural characteristic time 2τ) modified by a reducing factor $e^{-\alpha_0} \leq 1$.

Alfa (α) in equations (2) and (3) can be taken as a kind of 'growth potential', determinant of the increase (or decrease) of population in the time interval considered. Then,

$$r_b \times r_d = r_0^2 = (1/2\tau)^2, \text{ hence } 1/\tau = (r_b \times r_d)^{1/2} \tag{4}$$

$$r_b / r_d = e^{2\alpha}, \text{ hence } \alpha \equiv \frac{1}{2} \ln(r_b / r_d) \tag{5}$$

The population rate equation can be written therefore as

$$\frac{dN_2}{dt} = N_1 p_{12} - N_2 p_{21} \tag{6}$$

$$\frac{dN_1}{dt} = -N_1 p_{12} + N_2 p_{21} \tag{7}$$

Subtracting Eq. (7) from Eq. (6) we get

$$\frac{d(N_2 - N_1)}{dt} = (N_2 + N_1)(p_{12} - p_{21}) - (N_2 - N_1)(p_{12} + p_{21}) \tag{8}$$

which, taking into account that $\left[N_2(t) - N_1(t)\right] = \Delta P(t)$ is the increase in live population at time t, and $\left[N_2 + N_1\right] = \Delta N$ (constant), can be rewritten as

$$\frac{d\Delta P}{dt} = N 2 r_0 \sinh \alpha - \Delta P 2 r_0 \cosh \alpha \tag{9}$$

Using $p(t) = \Delta P(t)/N$, dimensionless, Eq. (9) becomes

$$\frac{dp(t)}{dt} = \frac{1}{\tau}\left[\sinh \alpha - p(t)\cosh \alpha\right] \tag{10}$$

The general solution of this linear differential equation (see Kreyszing, 1972) is

$$p(t) = e^{-\int (\cosh \alpha/\tau) dt} \left[\frac{\sinh \alpha}{\tau} e^{\int (\cosh \alpha/\tau) dt} dt + C\right] \tag{11}$$

In particular, for a step (up or down) in growth potential, say from $\alpha = 0$ at $t = 0$ to $\alpha \neq 0$ at $t > 0$, the integrals in Eq. (11) are straightforward, and we get

$$p(t) = e^{-(\cosh \alpha/\tau) dt} \left[\tanh \alpha \times e^{(\cosh \alpha/\tau) dt} + C\right] \tag{12}$$

which, from $p(0) = 0$ at $t = 0$ leads to

$$C = -\tanh \alpha \tag{13}$$

resulting in

$$p(t) = \tanh \alpha \left[1 - e^{-(\cosh \alpha/\tau) dt}\right] \tag{14}$$

Therefore, using $p(t) \equiv \Delta P(t)/N$ we finally get

$$P(t) = P_{RL} + \Delta P_{max} \tanh\alpha \left[1 - e^{-(t-t_i)/\tau^*} \right] \quad (15)$$

where $\tau^* = \tau / \cosh\alpha$, for a step up ($\alpha > 0$) in population.

Eq. (15) describes a step up in population due to an increase in fertility rate, a decrease in death rate (life expectancy increase) or a combination of both resulting in a net growth rate. The starting growth population level -occurring at time t_i- is denoted by P_{RL}. We note that only two numerical parameters (in addition to P_{RL}) are needed to describe the time evolution of the whole set of World population data for such the step up: the population growth potential α (related to the ratio of birth rate to death rate, as given in Eq. (5)) and τ^* (related to the product of the same rates as given in Eq. (4)).

As will be seen below, the model works: it fits well the set of UN data from 1950 to 2010 and suggests a declining future trend for the world population based upon UN data for 2000-2010. The rate equation model has the definite advantage of reflecting recent past changes in birth rates and death rates, which are determinants of the 'momentum' of world population evolution.

Analysis of World Population Data (1950-2010)

In order to analyze in detail the available UN world population data (1950-2050), and to make some qualitative considerations about future trends, it is convenient first to introduce an empirical relationship between the birth rate ($BR = r_b$), defined as the number of births per year per 100 population, and the fertility rate (FR), defined as the average total number of children per woman.

Fig. 2 gives the data for birth rate vs. fertility rate for China, India, USA and Russia (1995-2010) summarized in Table 1. The relationship

$$(BR) = 0.72(FR) \times 10^{-2} \qquad (16)$$

is very approximately fulfilled. Here the proportionality factor between (BR) and (FR) has been slightly corrected to take into account that the global female to male population in those countries is about. (% female)/(% male) $\cong 0.935$.

Table.1: Birth rate and fertility rate for various countries (1995-2010).

Year	China BR	FR	India BR	FR	USA BR	FR	Russia BR	FR
1995	1.82	2.3	2.93	3.9	1.57	2	1.07	1.6
2000	1.62	1.8	2.59	3.2	1.47	2	0.93	1.4
2005	1.3	1.7	2.33	2.9	1.41	2.1	0.96	1.3
2010	1.24	1.7	2.23	2.8	1.4	2.1	1.05	1.4

Source UN.

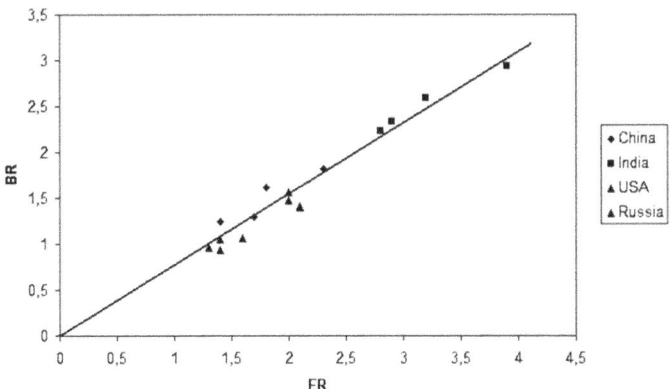

Figure 2: Birth rate per 100 population versus Fertility rate for China, India, USA and Russia (1995-2010). An excellent correlation is found by means of $(BR) = (0.72)(FR) \times 10^{-2}$.

(Source: www.unpopulation.org)

UN data for world fertility rates (FR) and the world growth rates (GR) for 1950-2010 are available at United Nations web, from which the birth rate ($BR = 0.72 \times FR$) and the death rate ($DR = BR - GR$) are directly obtained. The corresponding numbers for growth potential $\alpha = \ln(BR/DR)/2$ and inverse characteristic time $1/\tau = 2(BR \times DR)^{1/2} \times 10^{-2}$ are given in subsequent columns. Projected UN rates for subsequent years are given for completeness. It can be seen that, in spite of the fact that (BR) and (DR) vary considerably with time between 1950 and 2000, the numerical values for $x = BR/DR \approx 2.57$ and $\alpha = \ln(BR/DR)/2 \approx 0.471$ can be used as representative values for the half century through which a large step up in population is taking place. The numerical value for τ, however, varies smoothly from about $\tau = 18.4$ years in 1950 to about $\tau = 40.18$ years in 2000. This can be correlated to the increase in life expectancy and possibly to a marked delay in the life-giving age for women. (Table 2 gives in the first two columns the UN world fertility rates (*FR*) and the world growth rate (*GR*) for 1950-2010, from which the birth rate $BR = 0.72 \times FR$ and the death rate $DR = BR - GR$ are directly obtained.)

Table 2: UN demographic data (1950-2010) and projections (2010-2040).

Year	FR (UN)	GR (UN)	BR (Eq.16)	DR (Eq.17)	x BR/DR	α Eq.5	τ^{-1} Eq.4	τ^* $\tau/\cosh\alpha$
1950	5.02	1.80	3.61	1.81	1.99	0.345	0.0512	18.42
1960	4.97	1.97	3.58	1.61	2.22	0.400	0.0480	19.28
1970	4.48	1.94	3.23	1.29	2.51	0.460	0.0407	22.17
1980	3.57	1.71	2.57	0.86	2.99	0.547	0.0297	29.15
1990	3.03	1.50	2.18	0.68	3.20	0.582	0.0244	34.93
2000	2.48	1.07	1.79	0.72	2.50	0.457	0.0226	39.98
2005	2.22	0.87	1.60	0.73	2.19	0.393	0.0216	42.98
2010	2.05	0.74	1.48	0.74	2.01	0.348	0.0208	45.21
2020	1.87	0.48	1.35	0.87	1.55	0.220	0.0216	45.19
2030	1.71	0.17	1.23	1.06	1.16	0.074	0.0229	43.62
2040	1.59	0.10	1.14	1.04	1.10	0.046	0.0219	45.67

Using the data from the Population Division of the UN Department of Economic and Social Affairs we may analyze in detail the *step up* in world population taking place between 1950 and 2010, and then, taking into account that *x* (and α) begin to decrease in the year 2000, we can guess on the time for the incipient *step down* in world population.

Results

In Fig. 3, the normalized population $p(t) = \tanh\alpha \left[1 - e^{t/\tau^*}\right]$, $\tau^* = \tau/\cosh\alpha$, is given as a function of t/τ^* for various values of $x = r_b/r_d$, ($\alpha = 1/2\ln x$). It can be seen that a replacement level (*RL*) is achieved in all cases for $t/\tau^* > 3$, and that the population at the replacement level grows in proportion to $\tanh \alpha$.

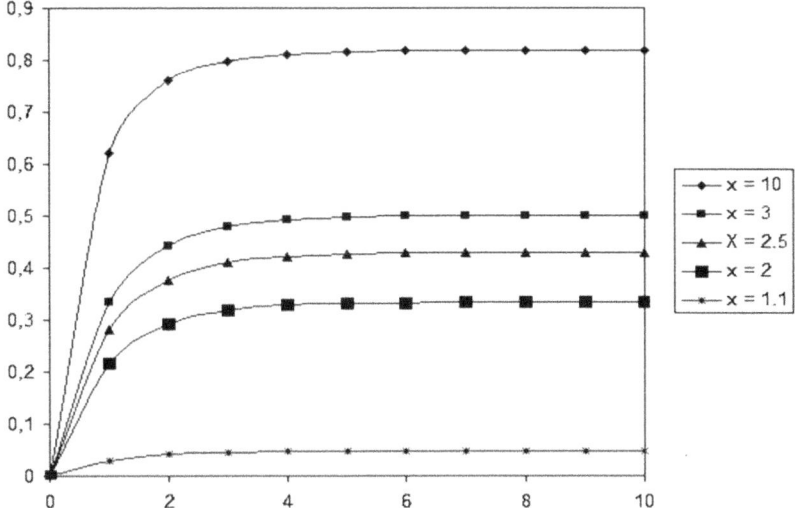

Figure 3: Normalized population increase versus normalized time for various $x = r_b/r_d$. Curves: $p(t) = \tanh\alpha\left[1-\exp(-t/\tau^*)\right]$; $\tau^* \approx \tau/\cosh\alpha$. (For different values of $x = r_b/r_d$, $\alpha = \ln x/2$.)

Figure 4 gives the birth rate (*BR*) and the death rate (*DR*) per year per 100 population as a function of time: actual UN data (1950-2010), UN (2004) projections for the period 2010-2050 are also given. It is seen that, in the interval (1950-2000), *BR* and *DR* decrease monotonously and smoothly, keeping the ratio (*BR*)/(*DR*) approximately constant. In the interval 2000-2010 a change in *DR* is taking place, probably related to the fact that in some countries, like Japan, the transient surplus population connected with the sustained increase in life expectancy, accompanied by the decrease in effective fertility rate, is beginning to fade away. During this interval the world population still increases slowly but it is leveling out. There is no such a thing as an exponential increase.

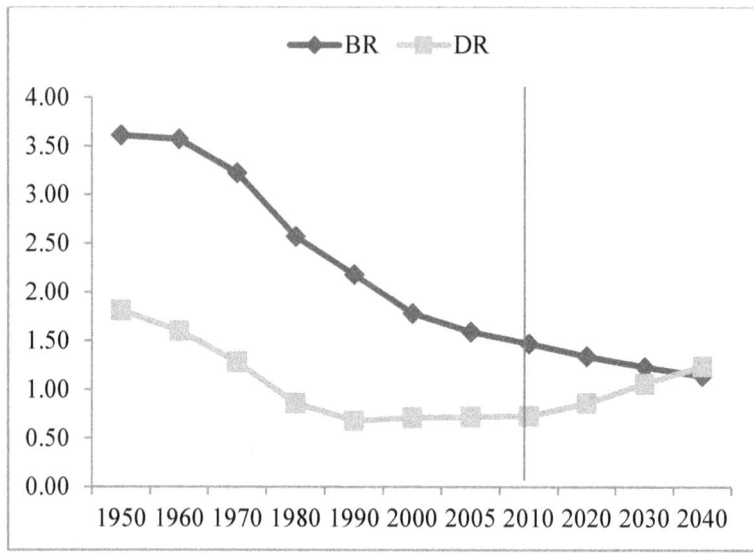

Figure 4: Birth rate ($\times 10^2$) and death rate ($\times 10^2$) UN actual data (1950-2010) and UN projections (2010-2050). See Table 2.

Figure 5 displays the *population growth potential* α (dimensionless) and the *characteristic time* τ (years) in the same time span. α remains practically constant up to $t = 2000$, and then begins to drop up to 2010. A linear extrapolation of the actual data for $\alpha(t)$ up to 2020 suggests a population decrease at about mid-century. On the other hand τ evolves smoothly from $\tau(1950) \approx 18.4$ years to $\tau(2000) \approx 40.2$ years and then begins to change tendency.

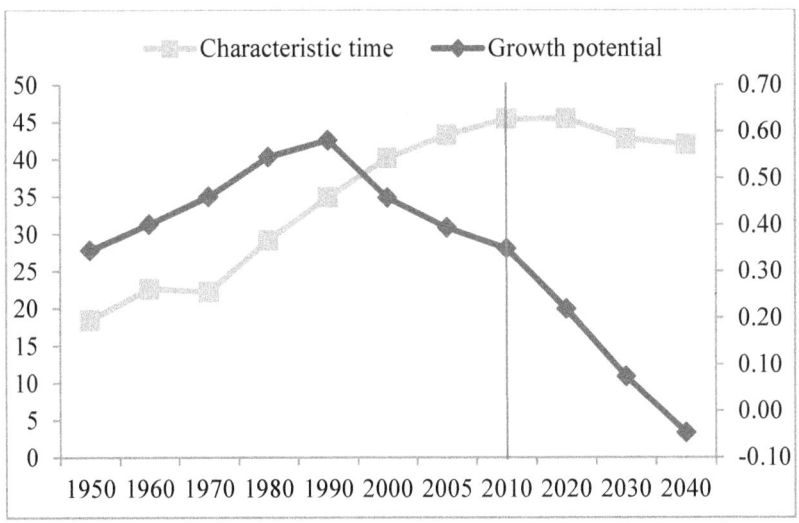

Figure 5: Growth potential (α) and characteristic time (τ): UN actual data (1950-2010) and UN projections (2010-2050). See Table 2.

If the characteristic time reflects a global tendency for women to have children at later ages (which might well be the case), by the year 2000, this age may be approaching already the age at which women become infertile. For the period 1950-2000 we can take an average value $\tau \approx 30.3$ years which would result in an effective $\tau^* = \tau / \cosh \alpha \approx 27.2$ years for the above period.

Finally Fig. 6 gives the actual UN data (and the 2004 projections) and the theoretical curves describing the time evolution of the world population (1950-…). The Malthusian projection[6] for population growth beyond 1980 is also given for comparison: $P(t) = P(t_0)(2)^{(t-t_0)/25}$ with $P(1980) = 4.5 \times 10^9$ at $t_0 = 1980$.

The analysis presented in this work could suggest further research in the factors influencing birth rates in women and men, as well as death rates. Regarding birth rates, of course there is a maximum number of children a woman wish or is able to have. Also regarding death rates there is a maximum which human nature is made to live.

Table 3: Population step amplitude in the period 1975-2010.

t	1975	1980	1985	1990	1995	2000	2005	2010
$t - t_i$	10	15	20	25	30	35	40	45
$\Delta P_m \tanh \alpha$	8.2	7.1	7.0	6.8	7.5	8.1	8.2	8.5

The step amplitude is obtained directly from the UN data for $P(t)$, $P_{RL} = 1.6$ and $\tau^*(t)$ for $t_i = 1965$ using:

$\Delta P_{max} \tanh \alpha \equiv \left[P(t) - P_{RL} \right] / \left[1 - e^{-(t-t_i)/\tau^*} \right]$; we take $\tau \sim \tau^*$ from Table 2.

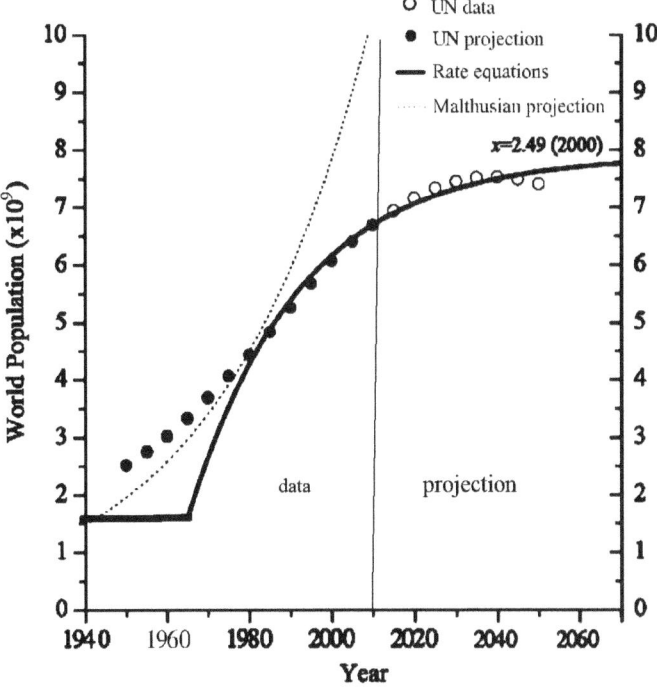

Figure 6: World Population $P(t)$ versus time. 1950-2010: UN data (red circles); Rate equations (continuous curve) 2010-2050: UN projections (white circles).

Concluding remarks

Our analysis of world population data by means of Eq. (15) fits the UN data satisfactorily and shows clearly that the population increase in 1950-2010 should be attributed more to the transient decrease in death rate level (related to the increase in life expectancy) than to nonexistent increase in birth rate, which was decreasing consistently already even before the 1950's, even before chemical contraceptives and legalized abortion begun to play any role.

An exponential population growth can be discarded as totally unrealistic at least at the next half century. Using fitting parameters extracted from recent UN population data, our rate equation solutions approach, which indicates that $x(t) = r_b(t)/r_d(t)$ may approach to one at $t \sim 2032 \pm 10$, suggests a decrease in world population beginning to take place about this time.

In 1960 population and world economic development was examined in *Science*. At about the same time, Foerster, Mora and Amiot (1960) reported in *Science* that that November 13th, 2026 would be the date at which world population would become infinity. The prediction was based upon an empirical equation for the population with a denominator going to zero as time increases. Today, 2011, world population is approaching 7.0×10^9 and, according to the UN data it will be around 7.3×10^9 in the year 2026 and approaching a maximum somewhat later.[7] (Reports in Science on the UN Conferences on World Population at Bucharest and Mexico were given by Boersma (1975) and Lutz, O'Neill and Scherbov (2003).)

The evidence for 'negative momentum' in Europe's population around 2000 is a result of low fertility rates. Caldwell (2008) most recently, criticizes Mathew Connelly's book 'Fatal Misconceptions. The Struggle to Control World Population'. Eberstadt (1997; 2001), on the other hand, points out that the world today may confront an unfamiliar crisis: rapidly decreasing birthrates and declining life spans that might set back the progress of human developments. Work by other authors share this perspective (see for instance Chaunu 1997; Ulrich 2000; Yea 2004).

The UN Press Release reports world population figures for 2050 and 2100, *if* fertility in all countries converge to replacement levels. Total fertility for the world and for countries grouped by fertility level in 1965-1970, 2005-2010, 2045-2050 and 2095-2100 are given. It may be noted that a large decrease between 4.8 and 2.5 is recorded for 1965-70/2005-10, much larger than the (expected) low decrease between 2.2 and 2.0 for 2045-50/2095-2100.

We think a rate equation approach is validated by the good fit of Eq. (15) to the UN data involving only two fitting parameters may provide a useful tool to simulate world population trends for the near future.

REFERENCES

[1] Population Division. http://www.un.org/esa/population/unpop.htm
[2] Asimov (1972: 795) gives a figure of 2×10^{16} Kg.
[3] They differ only by the speed of decrease for fertility. For more details see Leridon (2008).
[4] The partially similar approach has been developed by Miranda and Lima (2010; 2011). These authors employ logistic and power law methodologies for both retrospective and prospective analyses of extended time series describing evolutionary growth processes, in environments with finite resources. Their projections for world population are consistent with ours.
[5] We have estimated roughly the weigth per person as 60 ± 15

Kg. In any case this estimation does not change the general argument.

[6] As it is well known, T.R. Malthus held that "population, when unchecked, increases in geometric proportion ... doubling itself every twenty five years..." (Malthus, 1985[1798], Chp. I –II).

[7] The lower most likely estimate: The 2004 Revision/ The 2010 Revision. See esa.un.org/wpp/Documentations.htm (accessed March 26th, 2012).

FURTHER READING

Asimov, I. *Asimov's Guide to Science*. New York: Basic Books, 1972.

Boersma, D. "World Population Conference in Perspective." *Science* 188, no. 4193 (1975): 1069-69.

Caldwell, J.C. "Fatal Misconception - the Struggle to Control World Population." *Science* 321, no. 5892 (Aug 2008): 1043-43.

Chaunu, P. "From Explosion to Implosion of the Population - Vital Peril." *Cahiers d'Economie et Sociologie Rurales* 181, no. 9 (Dec 1997): 1923-33.

Dekker, A.J. *Solid State Physics*. Englewood Cliffs, NJ: Prentice Hall, 1962.

Eberstadt, N. "The Population Implosion." *Foreign Policy*, no. 123 (Mar-Apr 2001): 42-53.

"World Population Implosion?" *Public Interest*, no. 129 (Fal 1997): 3-22.

Foester, H v; Mora, P.M. and Amiot, L.W. "Doomsday: Friday, 13 November, A.D. 2026." *Science* 132 (1960): 1291.

Gonzalo, J.A. "On the usefulness of the hyperbolic functions to describe physical phenomena". *Ferroelectrics* 401 (2010): 9-16.

Gonzalo, J.A., Frutos, J. and Garcia, J. *Solid State Spectroscopies*. Singapore: World Scientific, 2002.

Groombridge, B., and Jenkins M.D. *World Atlas of Biodiversity. Prepared by the Unep World Conservation Monitoring Centre.* Berkeley, USA: University of California Press, 2002.

Kreyszing, E. *Advanced Engineering Mathematics*. New York: Wiley, 8th edition, 1998.

Leridon, H. "Human populations and climate: Lessons from the past and future scenarios." *Geoscience* 340 (2008): 663-669.

Loudon, R. *The Quantunm Theory of Light*. Oxford: Oxford Science Publication, 2000.

Lutz, W., B. C. O'Neill, and S. Scherbov. "Demographics: Europe's Population at a Turning Point." *Science* 299, no. 5615 (Mar 2003): 1991-92.

Lutz, W., Sanderson, W. and Scherbov, S. "The end of world population growth." *Nature* 412, (Aug 2001): 543-545.

Malthus, T.R. *Essay on the Principle of Population as it Affects The Future Improvement of Society*. London: Penguin Classics, 1985 [1798].

Miranda, L.C.M. and Lima, C.A.S. "On the logistic modeling and forecasting of evolutionary processes: Application to human population dynamics." *Technological Forecasting and Social Change*, 77(5) (2010): 699-711.

Miranda, L.C.M. and Lima, C.A.S. "On the forecasting of the challenging world future scenarios." *Technological Forecasting and Social Change*, 78(8) (2011): 1445-1470.

Ulrich, R.E. "Explosion of the World Population or Implosion?" *Internationale Politik* 55, no. 12 (Dec 2000): 17-24.

Vian Ortuño, A., ed. *Introducción a La Química Industrial*. Barcelona: Editorial Reverte, 1994.

Yea, S. "Are We Prepared for World Population Implosion?" *Futures* 36, no. 5 (Jun 2004): 583-601.

CHAPTER 3.3
ON WORLD POPULATION SLOWING DOWN[1]

In Science (4 November 1960) Friday 13 November AD 20026 was given as the "Doomsday" of planet Earth, a doomsday produced by "world population" going to infinity. In that paper, a rudimentary rate equation describing the evolution of world population with time was approximated in such a way that a quantitative calculation resulted in that "doomsday." In this paper, we give a more realistic rate equation respecting general conservation principles and compare previous results and the results of our calculation with actual UN data for 1960-2010 and UN medium term projections. At present there is disagreement among experts as to what is to be expected for world population in the years to come: some think population is still growing out control, some of them say it will be approximating a constant level about 2050 and others expect it to be in clear decline between

---

[1] Gonzalo, J.A., Muñoz, F.F., 2014. *Prospects of world population decline in the near future: a short note.* Departamento de Análisis Económico, U.A.M.

2050 and the end of 21st century. Our model shows that if no drastic and unexpected change takes place worldwide soon, world population will be slowing down at an accelerated pace after 2050. Our rate equation approach is similar to that used in condensed matter physics and chemical physics to describe the evolution of a two level system under an external perturbation and the results is much more realistic than a purely exponential result as generally assumed in the last decades of last century.

Introduction

'Doomsday: Friday, 13 November, A.D. 2026.' This was the original title of an article published in *Science* (4 November 1960) by Heinz von Foester, Patricia M. Mora and Lawrence W. Amiot.[1] They warned in the article's subtitle that at this date human population will approach infinity if it grows as it has grown in the last two millennia. Von Foester et al. begin arguing that in any biological system the time evolution of the total population is determined by two factors: fertility and mortality, and that the rate of change should be given by

$$\frac{dN}{dt} = \gamma_0 N - \theta_0 N = \alpha_0 N \qquad (1)$$

where $\alpha_0 = \gamma_0 - \theta_0$ may be called the 'productivity'. This equation results, of course, for $\alpha_0 = \text{constant}$, in an exponential (Malthusian) increase for any $\alpha_0 > 0$. The authors then consider the case of 'hypothetical paradise' in which no environmental hazards, no limited food supply and no detrimental interactions between the individual members of the population need to be taken into consideration. This is, of course, not very realistic, and they proceed to relax the assumption of $\alpha_0 =$ constant and, somewhat arbitrarily, they assume the productivity α changing as a function of N as

$$\alpha = \alpha_0 N^{1/K} \qquad (2)$$

where α_0 and K can be fitted to the available experimental data for human population in an extended time interval. Their approach leads to

$$N(t) = K/(t_0 - t)^n \qquad (3)$$

where k, t_0 and n are adjusted to the available world population data. k has the meaning of a population number at $(t_0 - t) = 1$ measured (arbitrarily) in years, t_0 is the so called 'doomsday time' and n is a dimensionless exponent that, using their chosen set of data, comes out close to one. Obviously, Eq. (3) blows up at $t = t_0$ for any $n > 0$.

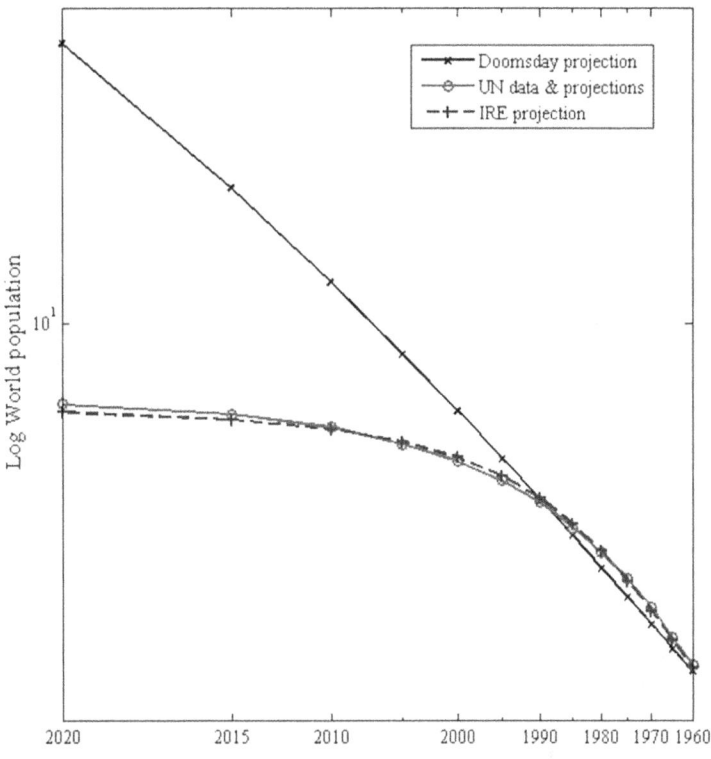

Figure 1: World population trends.

This is, of course, very unphysical. The total mass upon the Earth surface is strictly conserved. It cannot grow, and much less grow to infinity. The total biomass can grow somewhat, but always within definite limits. Further, von Foester et al. estimate K, t_0 and n for the human world population as $K = 1.79 \times 10^{11}$, $t_0 = 2026.87 \pm 5.5$ years (which can be approximated as 2026), and $n = 0.99$. Using these data in Eq. (3), it implies that

$N(1960) = 2.82 \times 10^9$ -somewhat lower that the UN estimate for that year, $N_{UN}(1960) \simeq 3.02 \times 10^9$. Using $K = 1.91 \times 10^{11}$ instead of $K = 1.79 \times 10^{11}$, takes care of the difference.

Figure 1 gives, in the same log-log representation used by von Foester et al. ($\log_{10} N(t)$ vs. $\log_{10}(k/(t_0 - t))$, i.e. the time evolution of total world population from well before 1960 to some years beyond 2010. It is seen that the UN data fall well upon the 'doomsday' curve from 1960 to 1990 but begins to deviate substantially thereafter. By 2010, $[P(t)]_{vF}$ as estimated by von Foester et al., is about 444 times larger than the actual $[P(t)]_{UN}$ value for that year.

In the final paragraph of their paper, the authors point out that among the suggestions made to solve the problem of the incoming of the world population explosion, legislation, heavy taxation of families with more than two children, tax deduction cancellations, etc., and even *space travel* had been proposed recently.[2] But, they add, no re-entry permit to Earth can be given to those flying away from our planet.

About forty years later Wolfgang Lutz, Sanderson, and Scherbov (2001), published 'The end of world population growth' in *Nature*, in which they conclude that this growth is likely to come to an end in the foreseeable future. They improve on earlier methods of probabilistic forecasting and show that there is around an 85 per cent chance that the world's population stops growing before the end of the century, a 60 per cent chance that it will not exceed 10 billion people before 2100, and some 15 per cent that it will be less at the end of the century that at its beginning.

They conclude also that there is a 20 per cent chance

that a peak of population would be reached by 2050. As we will see below, using an 'improved rate equations' (*IRE*) model, and the additional actual world population data for 2001-2013, we conclude that a maximum (not a peak) in world population around 2045 is likely with a likelihood substantially higher than 20 per cent. The decrease in population from the maximum value around 7.75 billion in 2045, will be likely to around 7.66 billion in 2090, with a substantial decrease of 90 million.

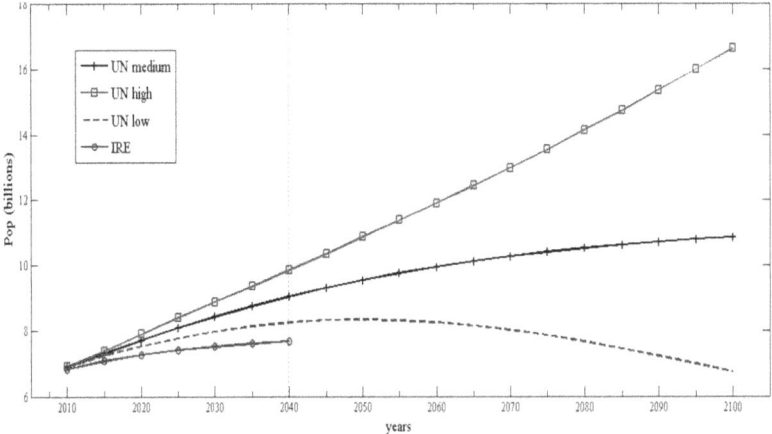

Figure 2: UN and IRE projections.

Figure 2 gives the estimated total world population in billions for the years 2000-2100 as given for the high, medium, and low fertility variant UN scenarios. The actual data for 2000-2010 (black circles) and the IRE projections (see below) for 2000-2040 (open circles) are also given. It may be seen that the low UN scenario fits better than the medium UN scenario the actual data for 2000-2010 and is closer to the IRE projection for 2000-2040.

Lutz et al. examine also regional population trends and conclude that the extent of the regional differences in the speed of population ageing, concomitant with population stabilization and decline, will pose major social and economic challenges. They conclude that the prospect of an end to population growth is welcome news for efforts towards a sustainable development, but this conclusion is far from substantiated for the time being as the recent global economic crisis clearly suggests. They say also that the main determinant of the timing of the 'pick' in population size is the assumed speed of fertility decline in the past of the world that still have relatively higher fertility. However, it may be noted that the spectacular increase in world population in the second half of the 20th century was due mainly to a sustained increase in *life expectancy* rather than to an increase in fertility, which did not ceased to go down from the early fifties onwards.

An improved rate equations approach

Recently, Gonzalo et al. (2013) proposed a novel rate equations approach to modelling world population trends. This approach is designed to describe *steps* up or down in population in a *two level system* (individuals in level (2) *alive*; individuals in level (1) *dead* or *potentially alive* in the relative abundant available biomass). The rate equations are:

$$\frac{dN_2}{dt} = N_1 p_{12} - N_2 p_{21} \qquad (4)$$

$$\frac{dN_1}{dt} = -N_1 p_{12} + N_2 p_{21} \qquad (5)$$

resulting in

$$\frac{d(N_2 - N_1)}{dt} = (N_1 + N_2)(p_{12} - p_{21}) - (N_2 - N_1)(p_{12} + p_{21}) \quad (6)$$

where $(N_1 + N_2) = N$ is related to the maximum jump in population, $\Delta P_{max} \times (N_2 - N_1)$ is related to the actual change in population $\Delta P(t)$ at time t, at which $N_2 = N_2(t)$ and $N_1 = N_1(t)$, and p_{12} is directly related to the birth rate (r_b), and p_{21} to the death rate (r_d). Two important dynamic parameters, $\alpha = \frac{1}{2} Ln(r_b/r_d)$ (dimensionless) and $\tau = (r_b \cdot r_d)^{-\frac{1}{2}}$ (a dimension of time) are defined. A general solution of Eq. (6) leads to

$$P(t) = P_{RL} + [\Delta P_{max} \cdot \tanh \alpha]\left[1 - e^{-(t-t_i)/\tau}\right] \quad (7)$$

P_{RL} is the population at replacement level before the jump, t_i is the inflection time for a jump in population (we are considering here a jump up) and τ^* is a characteristic time, obviously related to the effective fertility time span in the women population making up always about one half of the total human population.

Eq. (7) can be conveniently rewritten[3] taking into account that for a well defined jump in population $P(t)$ first grows gradually from P_{RL} towards $P_{RL} + \frac{1}{2}[\Delta P_{max} \cdot \tanh \alpha]$ and then grows gradually from this population to $P_{RL} + [\Delta P_{max} \cdot \tanh \alpha]$, as given by

$$P(t) = P_{RL} + \frac{1}{2}[\Delta P_{max} \cdot \tanh \alpha]\left[1 + \tanh \frac{t - t_i^*}{\tau^*}\right] \quad (8)$$

A good fit to the UN data is obtained with $P_{RL} = 2$ billion, $\Delta P_{max} \cdot \tanh \alpha = 2.93$ billion, $t_i^* = 1985$ and $\tau^* = 30$ years. This is what we have called the improved rate equations (*IRE*) solution for our model, describing in this case the time evolution of world population for a population jump up.

Projections and discussion

Before using Eq. (8) to analyse population data and to make momentum-like projections for the near future, let us investigate the trend of the annual relative increase in world population $\delta P(t)/P(t)$ (per cent per year) as a function of time. Figure 3 gives the actual UN population increase rate for 1950-2010 for 2004 and the revised UN population increased rate for 2013 as well as the increase rate obtained using Eq. (8).

It can be seen that extrapolating linearly from 1990 forward the UN data, $\delta P(t)/P(t)$ becomes zero about $t = 2045 \pm 16$. At this time world population must go through a transient maximum of about 7.75 billion. Thereafter if the birth rate continues to decrease, and the death rate (concomitant with a decrease in the average life expectancy due to overall ageing of the population as a whole) decreases also, the overall decrease in world population becomes something to be anticipated.

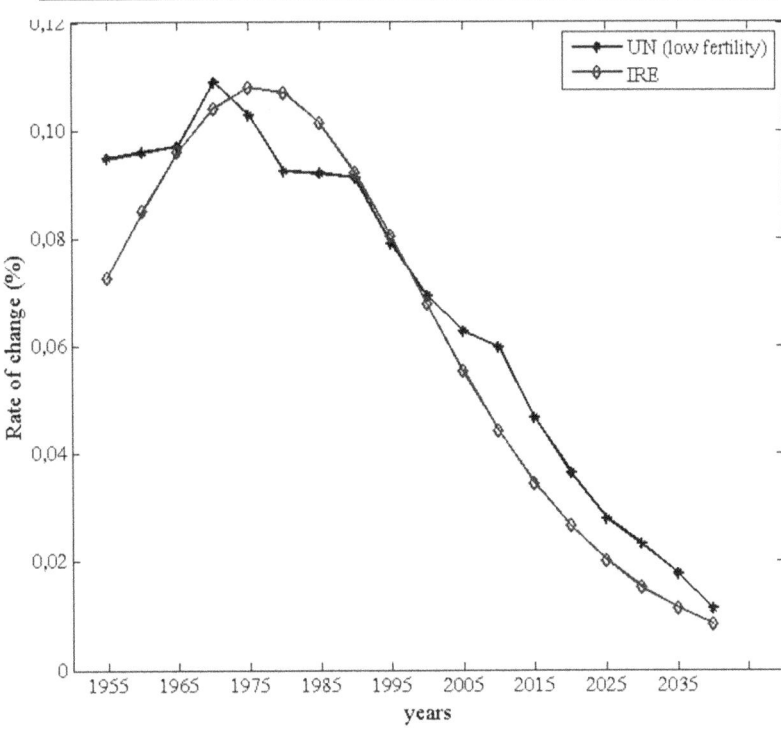

Figure 3: Rates of change of population: UN & IRE

Finally, we give in Figure 4 the UN world population data (black circles) for 1950-2010 and the world population projections obtained by means of Eq. (8), the IRE solution (open circles), for 2020-2045. We can see that the IRE solution fits very well the data. We show also in this graph that UN estimation projections for the world population in 2010 made in the period from 1990 to 1998 where substantially larger than the actual value at 2010, with the estimates decreasing gradually as they approached 2010. The UN overestimations of future world population for 2020 and 2030 in previous decades seem to be negligible if $P(t)$ calculated using Eq. (8) is correct.

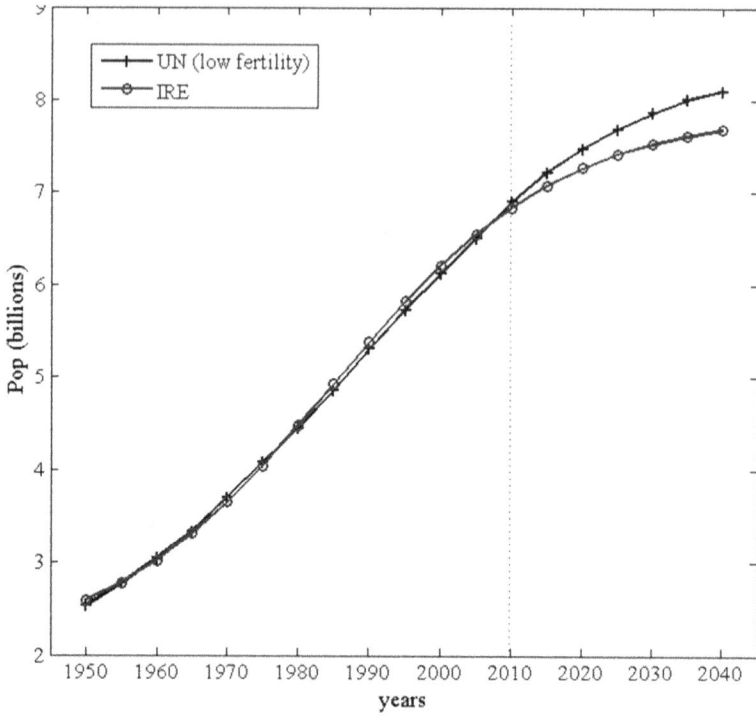

Figure 4: UN & IRE projections

Concluding remarks

Of course, no statistical model can predict changes in global socio-cultural trends, such as global increases in casualties of epidemics, or such as the rise or fall of world hegemony, like the Soviet Union in the past or perhaps China or the U.S. of America in the future, or such as European legislation leading to massive contraception (or possible reversals in these politics), as well as possible changes in demographic trends, favouring a voluntary fertility rate compatible with a slow increase in population. But in the short and medium term projections based upon our *IRE* model, which take properly into account the role of momentum, may be used to demand attention about the incoming slow down (and eventually decline) population in the second half of this century, and may prepare the ground among economists, sociologists, and politicians to face the inevitable consequences.

REFERENCES

[1] Foester, H. v., Mora, P. M., & Amiot, L. W. (1960). Doomsday: Friday, 13 November, A.D. 2026. *Science, 132*, 1291
[2] Christopher, R. C., & Griffith, T. (1960). Time, 75, 22.
[3] Gonzalo, J. A., Muñoz, F.-F., & Santos, D. J. (2013). Using a rate equations approach to model World population trends. Simulation: Transactions of the Society for Modeling and Simulation International, 89(2), 192-198. doi: 10.1177/0037549712463736

Stanley L. Jaki (1924 – 2009)

CHAPTER 3.4
AT WHICH SIDE WILL BE HISTORY AT THE END OF THE 21ST CENTURY?

According to **Stanley L. Jaki**, the great Hungarian-American historian of science, "history", i.e. temporary or permanent victory, will be at the side of those who have the capacity to put into practice the most apt instruments to develop a more advanced society, a society capable of putting itself well above the rest. This was true in the past regarding local conflicts, but it became more evident globally in the twentieth century, when the development of the most apt instruments became clearly connected with advanced technologies dependent of the most exact sciences.

Herodotus, **Xenophon** and **Polybius**, the pioneers of historiography who were consummate analysts of the past in classical times, did not ask themselves whether history had a direction or a particular meaning. In particular, when the latter faced the conquest of Greece by Rome (a nation of barbarians for Polybius) he put history for the moment on the side of the Romans, but he pointed out that, in the end, the Romans will fall down themselves because no society, no army is above the **law of history**, which for Polybius, as for all ancient Greeks, was the law of the wheel or swastika the inevitable sequence of birth, progress, decadence and death, alternating again and again one after the other.

In a Christian perspective, history will be, at the end, on the side of God. It is increasingly evident however that the Christian foundations of Western Civilization are already under attack since the middle of the Eighteenth Century, specially so in Europe and America since the last third of the Twentieth Century. All Pagan Civilizations, as clearly illustrated by Stanley Jaki, share the belief in an eternal world in which everything repeats itself without end. The Judeo Christian tradition on the other hand affirms from the very beginning an **origin** and an **end** for the world. Modern **science**, the only viable science, has arrived in the 20th century to the concept of Big Bang Cosmos which implies an origin and an end of space-time in a finite universe. Of course there are contemporary cosmologists who tray to escape the inescapable implications of this, which entail a contingent, finite, created universe. But they are far from capable of being able to do it in a consistent fashion (logically and metaphysically).

Marxism / Leninism = Dialectic **Materialism, Darwinism** = Evolutionary **Materialism, Nazism / Fascism** = Racist / Deterministic **Materialism**, have dominated the cultural landscape of Europe and the world during most of the Twentieth Century.

The three are materialist, the three deny an immanent-transcendent Creator of the world, and the three dismiss a spiritual dimension in man. **Soviet Communism**, defeated after a long cold war, has left a bloody heritage of one hundred million victims, Materialist **Darwinists**, assuming that the Earth is overpopulated, through contraceptive pills and abortions are now suppressing many millions of human lives every year, and **Nazism** did suppress sixteen million innocent lives, in a decade (1934-1944).

The **"progress"** promoted by Marxists, Darwinists and national socialists has been (and still is) a very **bloody "progress"**.

How can be called "progress" the trivialization of provoked abortion? How can be seen as "progress" the fact that more than one third of the youth today consumes drogues? How can be considered **normal** sex *"contra natura"*? Many innocent children come to the age of reason today in a world without moral reference of good and evil. A world in which everything seems to be valid. A world in which a man, like a fish, seems to be guided only by instinct: the large fish devours the small fish. A world in which right and wrong are not objective categories but subjective concepts used to contradict and insult the adversary.

Since the Christian medieval epoque till relatively recent times Classical Greco-Roman wisdom and the Mosaic Decalogue have been the reference frame of the basic legislation, the uses and the costumes in Europe and America. In particular, as pointed out by the American Monk Fr. **M. Raymond, O.C.S.O.**, religious freedom landed in America bringing persecuted English Catholics in the "Ark" and the "Dove" before the "Mayflower" brought persecuted Puritans. American democratic freedom as registered in the Constitution does not proceed originally from Thomas Jefferson. It proceeds from earlier medieval scholastics, such as the Jesuits Francisco Suarez and Cardinal Belarmino and the Dominican Francisco Vitoria, professor of Philosophy at Paris and at Salamanca, who was the father of International Law. The building of the United Nations in New York did recognize Vitoria's contribution dedicating to him a special room.

In the first decades of the 21th century things are changing radically. The United nations, the O.I.T. (International Organization of Work), the International Red Cross, to mention only a few international organizations of the highest rank, declare themselves "formally neutral" but are in fact clearly hostile to the Christian way of life which is based on the natural rights as inscribed by the Creator in human conscience and encapsulated in the Ten Commandments. The large billionaire American and European foundations, for reasons not too clear to the ordinary observer, are not neutral either. They seem to favor progressive criteria (in fact, agnostic and atheistic criteria) hostile to the Christian way of life. Today fiscal exemptions to the humanitarian, medical and charitative help provided by Orthodox Christian organizations (including the Catholic Church) are conditioned and reduced as much as possible, while pro-abortion international organizations, like "Planned Parenthood" (which not only is against responsible motherhood and fatherhood but increasingly active in making money (a lot of money) selling the byproducts of massive abortions, are given massive support and protection by the UN and other international organizations.

People has not become well aware yet that the considerable increase in world population which took place in the second half of the 20th century was a consequence of the increase in **life expectancy**, associated to a considerable decrease in the rate of mortality, and not to an increase in the women **fertility**, totally inexistent in 1950-2000.

Forty years from now, by the middle of the 21th century (demographic tendencies have considerable inertia) **world population** is expected to **decrease** with unwelcome consequences, to say the least, for the world at large.

Careful analysis of the evolution of world population in terms of **birth rate** (BR) and **death rate** (DR) and the evolution of available **food, water** and **energy** show that world population could be at least twenty or thirty times the present population of seven billion people (see for instance in world population: Past, Present and Future (World Scientific 2016). In fact food production due to the "green revolution" has increased much more than population worldwide in 1950-2000.

Christopher Dawson (1889-1970) the great British historian, unequaled in the 20th century as an historian of culture, pointed out that Renaissance **humanism** can be regarded as the fountainhead of the noblest aesthetic and ethical ideals of civilized man. Such ideals have been besieged for many years by a ruthless materialism which has been degrading progressively the human spirit. At the beginning of the 21st Century, a couple of decades after the fall of Soviet Communism in Russia, the degradation is becoming more and more evident. In "Christianity and the New Age" (Reprinted by Sophia Institute Press: Manchester, New Hampshire, 1985) Dawson tells why humanism did succumb to the inexorable force of its own inner contradictions. The real source of its transcendent values was obviously Christianity, which alone possesses still the spiritual dynamism necessary to maintain those values alive.

Men of the Renaissance rejected their dependence on the supernatural and vindicated their supremacy in the temporal order. The self-affirmation of man leads gradually to the denial of the spiritual foundations of his freedom. Freedom to know intellectually beyond the level of the senses, and freedom to love personally beyond the level of the instincts. In **philosophy** this leads to the rationalism of Descartes, to the empiricism of Locke, to the radical skepticism of Hume, and from there to the rank subjectivism of Kant, Fichte and Hegel.

In **science**, the growth of man's control over nature was accompanied by a growing sense of human dependence on the material forces. Science, however, is nothing else than the intellectual penetration of scientists (a few geniuses and large number of ordinary researchers) on the ordinary laws of nature, and it is therefore not "materialistic" in itself. Science, as a human endeavor, is the result of using the human intellect freely, laboriously and successfully, to discover nature's secrets. Man's intellect, man's will and man's freedom are therefore involved in doing science.

In **economics** the independence from the supernatural and the supremacy of the temporal, leading to materialism, ends in the subjection of man to the rule of the machine, i.e., in the mechanization of human life. At the end, economic progress will be viewed as nothing else than the struggle among classes propounded by Marx and Lenin. In the aftermath of the Second World War, however, social security and ordered competition **under freedom**, demonstrated abundantly that precisely free ordered competition combined with social security is much better for ordinary workers both in Western Europe and America, than the war of classes.

In the **political** and **social** sphere, on the other hand, the **rejection** of the medieval principle of **hierarchy**, and the unlimited increase of secular power, lead to the **absolutism** of the modern national state, capable of going much further in practice than a feudalism tempered by workers guilds, municipal representation and the cultural influence of free universities, still ruled by Christian Catholic principles.

Behind all **religion** and **spiritual philosophy** there is a metaphysical **first step**, the affirmation of Being. Behind **materialism**, on the other hand, there is only a metaphysical negation, the denial of Being. As a young and competent Spanish lawyer (J.A. Primo de Rivera) said, only two years before Marxist-Leninists bullets ended with his life in November 1936, justice and truth ceased to be objective categories of reason when Jean Jacques Rousseau declared them to be so in the "Social Contract" (1762).

Let us take a look to the socio political trends discernible today in North America (USA), Europe, China, Russia and the Islamic World. Socio-political trends can change of course but in the long run fatal determinism is never granted in human affairs. But significant trends, however, if they go on unchecked for a long time must be taken seriously.

Table I gives the trend in Population (10^9) and in Gross Income ($\$ 10^{12}$) for the World, USA, Europe, China, Islam and Russia for 1992-2016.

Table II gives this umber of Christians, Muslims and Atheist in the World for 1992-2016.

As noted by William Kilpatrick in "Christianity, Islam and Atheism" in today's **America,** and more so in today's **Europe**, **"Multiculturalism"** is the official policy of the State, a kind of quasi-religious establishment, hostile to the classical Christian heritage based upon Greco-Roman wisdom, the Old Testament , the four Gospels and the New Testament.

To recover freedom in America, Europe and the World at large, relativistic "multiculturalism" must be defeated: There is no other way out.

Table I. Global demographic & economic trends 1992-2016
Population (10^9) Gross Income ($ 10^{12}) (*)

Year		1992	2000	2008	2016
World	Population	5.39	5.97	6.58	7.28
	Gross Inc.	21.2	30.1	47.7	78.2
USA	Population	0.25	0.25	0.30	0.32
	Gross Inc.	5.2	7.78	13.1	17.6
Europe	Population	0.38	0.37	0.46	0.47
	Gross Inc.	5.8	8.5	14.3	17.7
China	Population	1.36	1.25	1.55	1.37
	Gross Inc.	0.4	1.0	2.6	11.1
Islam(**)	Population	(0.95)	(1.15)	(1.38)	(1.70)
	Gross Inc.	-	-	-	-
Russia	Population	0.29	0.14	0.14	0.14
	Gross Inc.	2.6	3.9	0.8	1.9

(*) Data taken form Encyclopedia Britannica, Book of the Year
(**) Data for Islam from Global Data on Religious Adherents (E.B.)
Figures are rounded and may be shifted one or more years back.

Table II. Christians, Muslims and Atheists in the World 1992-2016 [*]

Year	1992	2000	2008	2016
Christians (10^9)	1.78	1.97	2.20	2.41
Muslims (10^9)	0.95	1.15	1.38	1.70
Atheists (10^9)	0.23	0.15	0.15	0.14

(*) Data taken form Encyclopedia Britannica, Book of the Year. Figures are rounded and may be shifted one or more years back.

www.ingramcontent.com/pod-product-compliance
Lightning Source LLC
Chambersburg PA
CBHW052322220526
45472CB00001B/231